SIX SHORT ESSAYS FOR UNDERSTANDING HISTORY

Cristiano Rimessi

INDEX

Preface — 4

The explorer as hero — 8

Traveling in the service of science — 30

The non-economic reason for making Libya a colony — 42

The postcolonial 'crisis' in the national building of Algeria, Syria and Iraq — 52

The European integration process for maintaining peace: successes and failures — 65

What went wrong with Turkey integration in the European Union — 93

Conclusions — 99

PREFACE

The main purpose of this collection of briefs essays is to address multiple issues in a very synthetic way in order to give to the reader a quick but valid overview over big historic problems. Six different papers are gathered in this book with not any particular link among them. This means that each dissertation is readable and understandable by itself.

However, there are two main themes that emerge from the two sections of this publication.

The first one concerns colonial and post-colonial history, and the second one involves European integration history. Both topics are related to European history, which constitutes the lens used for seeing and understanding the world. This does not mean that the method adopted is Eurocentric. On the contrary, there will be strong criticism of the European role in history without ignoring its merits.

The book starts with a sophisticated analysis of the explorer's role between the end of the 18th century and the first half of the 20th in British society. The reading key and the research focus are about the explorers' role and his depiction as a hero celebrated and honoured by European institutions as a role model, a fearless adventurer that opened the old continent to a majestic bigger world.

After all, this is the core of the history of exploration.

A progressive phenomenon that allowed Europeans to achieve a better knowledge and awareness of the limits of their physic world in an epoch in which travelling was life-risking, complex, dangerous but also exciting and incredibly fascinating. It will explain why the explorer became a symbol of this enterprise and what it meant to be such an icon in that period.

The second essay is strictly linked to the first one.

However, rather than a human being, this time the protagonist is a discipline born out of Western's mind. The research will deal with the role of science in the exploration's effort, arguing that the process of accumulation of knowledge did not simply helped Europeans to fill gaps in the maps or to classify animals, plants and other human beings. Science also paved the way for a more violent and aggressive process known as the colonization, a project that could take form only thanks to an unprecedented effort in expanding European awareness of the world outside their borders. Additionally, it will be also taken into consideration the experience of one of the first explorers who pushed himself in the interior of Africa: Mungo Park.

The third research take a step forward in history, when the times of exploration eventually ended and Europeans finally achieved their real purpose: the colonization and military invasion of extra-Europeans lands.

Generally speaking, the first thing that come to mind regarding the colonies is their economic advantages for the motherland. Frequently, colonial powers invaded other territories for commercial interests or for a systemic exploitation.

This is not the case for Italian subjugation of Libya, a country that had little to offer. It will be argued that there were little purposes for conquering such an inhospitable land. What Italians were really looking for was a *place in the sun,* glory, respect from other European powers and *grandeur* for their young nation.

The fourth dissertation deals briefly with the colonial legacy in Algerian, Syrian and Iraqi context, underlying common aspects and differences. The main argument is about the colonial responsibilities in the long path of nation-building that these states had difficultly taken. During the exposure it will be contested also the theory that nations in the Middle East and North Africa were created by Europeans only by drawing lines in the sand because the population living there do not have any national identity. On the contrary, it will be showed evidence of true and genuine nationalism coming from those populations.

The fifth paper is about the European integration process. After the Second World War, the continent is devastated.

Centuries of conflicts, political and military expansionism, exploitation of territories from all over the world and violence led to the most impressive conflict in history. From the ashes a new continent is eager to born. The most belligerent super-powers of history that risked to destroy the world are now willing to cooperate in a completely new framework in order to guarantee peace at least over their continent. The key for such objective is to bind themselves in some economic pacts. Unfortunately, this method has some important limits that are soon rise to surface by a war right in the corner. The process of European integration did little to prevent the Balkans conflict in the 90s as it will be demonstrated.

The final essay takes again the European integration process and stretch it to its natural limits. If the book is opened by a paper regarding the exploration of the extra-European world outside its borders, the last essay calls into question the inner identity of Europe by analyzing if the modern-day Turkey could really be considered part of the European culture.

Even though Turkey is hardly considered a democracy now days and as a consequence its integration in the European Union would not be appropriate, its culture cannot be accused of lacking of 'Europeannes'.

THE EXPLORER AS HERO

Since the most ancient time, human being has always been fascinated by two things: storytelling and role models. Frequently, these two elements are combined and the great gestures of a leading figure as for example a mythical hero, are narrated by someone or by the hero himself.

The point of this research paper is firstly to analyze in which way the explorer-hero has been taken by his nation as a propagandistic element underlying and explaining also why this gave to the nation and the society prestige and a reason to be proud of. Secondly, it will be clarified for what extent the explorer-hero had an impact over the public and the culture, considering both his époque and the following one. In doing so, it won't be ignored the newspapers' role in narrating the hero's feats but also in general, all the cultural production such as books, novels, articles and so on. Thirdly, it will be pointed out what characteristics make possible to recognize an explorer of those times as a hero, without ignoring the fact that in some cases, there were also dark aspects. Fourthly, it will be showed that these heroes influenced the society in which they lived and then, the same society contributed in the depiction of their heroes by narrating their feats. This testifies a relationship of interdependence and reciprocity.

This research paper takes into consideration four explorers: Livingstone, Stanley, Cook and Scott. They all lived during the British expansionism and imperialistic period, but they belong to different generations.

Between the end of the 18th century and the first half of the 20th the explorer achieved his maximum celebration and still have some impact today. As a demonstration of that, in 2002 the BBC broadcasted a television poll in which the spectators could vote the greatest British people in history. The result was quite impressive because among pop celebrities, politicians, kings and queens there were at least four explorers: Shackleton, Cook, Scott and Livingstone.[1]

In order to have an idea of what kind of impact an explorer had on his own society, David Livingstone could be a remarkable example. He started exploring the interior of Africa when he was 27 in 1840 until his death in 1873. In 1856 he crossed Africa from the west to the east coast while in the other two expeditions he tried to find the source of the Nile and habitable land for white colonies. Livingstone could be seriously considered the perfect prototype of explorer and role model.

---

[1] Matt Wells, 'The 100 greatest Britons: lots of pop, not so much circumstances', *The Guardian,* 22 August 2002

He embodied all imperial virtues of the Victorian age, such as fortitude, faith in God and resilience while his exploration concerned both imperialistic and scientific purposes.[2]

However, what made Livingstone particularly famous is an event linked with another explorer much more controversial: Henry Morton Stanley. In October 1871, Livingstone was in Ujiji, near the lake Tanganyika and he was incapable of going on with his exploration because gravely ill while traveling hired by the Royal Geographical Society.[3]

In that moment, Stanley, the correspondent of the New York Herald, arrived in the same place, and when he first met Livingstone, after having doffed his helmet, he bowed, and he pronounced the famous sentence: <<Dr. Livingstone, I presume.>>[4]

This sentence had an incredible impact over the public at that time because it became widely known, while his image doing the greeting gesture was reproduced in a lot of advertisements.[5]

---

[2] Elleke Boehmer (eds), *Empire Writing, an anthology of colonial literature 1870-1918* (New York, 1998) p, 487
[3] Elleke Boehmer (eds), *Empire Writing*, pp. 40-41
[4] Elleke Boehmer (eds), *Empire Writing*, pp.49-50
[5] Felix Driver, *Geography Militant: Cultures of exploration and empire* (Oxford: Blackwell, 2001), pp. 121-122

Additionally, the sentence is now the title of a song by the British band *The Moody Blues*[6] and the meeting between the two explorers inspired also a movie, *Stanley and Livingstone* by Henry King and Otto Brower.

The reason why this encounter left a mark in culture is probably because Stanley and Livingstone meeting has been narrated not only by the letters and books, they wrote about it, but also because newspapers reported the event extensively. The New York Herald in fact, published the letter written down by Livingstone in the article of 26 July 1872. In the same article, Livingstone is celebrated for some specific characteristics. His braveness, his scientific interest in exploring Africa and discovering the source of the Nile, his spirit of Christianity and his willingness in stopping the slave trade in Africa, a topic of great relevance in the United States as slavery had been abolished only few years before.[7] Livingstone is defined by Stanley as a hero and as a man of iron[8], high-spirited, brave, impetuous, and enthusiastic man. Like many else, Stanley admired Livingstone's abiding faith in God and his remarkable work as missionary in the African continent.[9]

---

[6] Moody Blues, *Dr. Livingstone, I Presume*, available at https://www.youtube.com/watch?v=TPkc6VOsqZ0, [Accessed 12/04/2020]
[7] An, 'John's Livingstone Testimony', *The New York Herald*, September 09 1872, p.3 & An, 'The Slave Trade in Eastern Africa', *The New York Herald*, August 13 1872, p. 4
[8] An, 'Livingstone', *The New York Herald*, July 15 1872, p.5
[9] A. W. Greely, *Explorers and Travelers* (New York, 1894)

What is more, he is regarded as a hero also because his struggles have been determined by betrayals against him by the Arabs who worked and traveled with him. Having enemies and opponents is a relevant component of the hero's narration, something that makes easier to construct this image.[10]

By all the aforementioned qualities we can understand what kind of perception there was about Livingstone.

Moreover, the New York Herald supported and celebrated both Stanley and Livingstone because Stanley's achievement in finding the lost explorer would have been seen as a journalistic triumph.[11] Another article published on the second of July had already reconstructed accurately Livingstone's journey from his departure from England to the interior of Africa near lake Tanganyika and his meeting with Stanley. The reporter-explorer was economically sustained for finding the lost explorer in Africa by the New York Herald. Finding Livingstone, in fact, was something of great interest at that time.

Judging by the amount of space given to the narration of Livingstone's adventure, it would not be wrong to assume that the readers of the New York Herald were quite interested in knowing the details of the events that occurred during an exotic exploration of a foreign and far land such the interior of Africa.

---

[10] An, 'Livingstone', *The New York Herald,* July 15 1872, p.5
[11] An, 'Doctor Livingstone, Grand Triumph for England and America', *The New York Herald,* July 26 1872, p.6

The narration itself of Livingstone misadventures, contributed to the depiction of an extraordinary man who could deal with extraordinary situations and adversities.[12]

However, Livingstone's celebration did not overshadowed Stanley.

On the article published on the 15th of July he reconstructs properly his voyage and his challenges in finding Livingstone. In addition, the herald expedition led by Stanley was <<as if guided by the hand of Providence, nor a month too late nor a month too soon>>[13] in helping Livingstone, sick and betrayed by the Arabs. In the conclusion of the same article, Stanley underlines the multiples way in which he helped the re-found explorer while in the article of 9th September, the journalist and explorer received congratulations from some American personalities.[14]

Despite the great endorsement Stanley received, there were also many critics on his account. Most of them came from two newspapers: *The Saturday Review* and *The Spectator,* from The Royal Geographical Society and the philanthropic community.

---

[12] An, 'Livingstone Herald Special from Central Africa', *The New York Herald*, July 2 1872
[13] An, 'Livingstone', *The New York Herald,* July 15 1872, p.5
[14] An, 'John's Livingstone Testimony', *The New York Herald,* September 09 1872, p.3

The New York Herald's correspondent has been accused of fraud, sensationalism and of not having offered an extensive geographical knowledge by exploring the African continent.[15]

Even though Stanley was a controversial figure in England, in the United States he was generally celebrated as a role model. In 1894 he was already included in a book recollecting all great American explorers: *Explorers and Travelers* written by General A. W. Greely.

Defined as a man who << stands forth a self-made man to whom strength has been accorded to develop the manhood that God implanted in his soul>>[16], Stanley was also positively considered by Gambetta, a French statesman who stated to him:

<<Not only have you opened up a new continent to our view, but you have given influence to scientific and philanthropic enterprises which will have its effect on the progress of the world. What you have done has interested governments—proverbially so difficult to move—and the impulse you have imparted, I am convinced, will go on year after year.>>[17]

In fact, Stanley significantly contributed to the establishment of the Congo Free State not only by exploring the territory but also by founding and establishing scientific and military stations.

---

[15] An, 'John's Livingstone Testimony', *The New York Herald*, September 09 1872, p.3
[16] A. W. Greely, *Explorers and Travelers*
[17] A. W. Greely, *Explorers and Travelers*

It has been already said that Stanley was a much more controversial personality compared to Livingstone. It could be argued that while Livingstone reflected a more scientific and religious hero model, Stanley even if invaluably contributed to scientific discoveries such as the source of the Nile, was much darker because some of his expeditions were to claim the highest fatalities of any recorded African expeditions.[18] This attracted many critics especially from the Exeter Hall, the philanthropic community. Furthermore, John Scott Keltie, a geographer of the RGS, considered his expeditions as half way between military campaign and heroic adventure.[19] Stanley was viewed also as the embodiment of a new way of exploring based on warfare and sensationalism in the way in which his feats were narrated.[20] When in 1885 the Berlin Conference was called, Stanley represented Belgian King Leopold's interests which is quite relevant because give us back the idea of how much important an explorer at that time could become. What is more, at that conference participated fourteen European countries and the United States which would be overestimated to assert that they participated thanks to Stanley's relevance but certainly to have an American citizen among the main personalities of the conference helped that nation in putting itself in a position of prestige.

---

[18] Elleke Boehmer (eds), *Empire Writing*, p. 495
[19] Felix Driver, *Exploration by Warfare*, p. 121
[20] Felix Driver, *Exploration by Warfare*, p.123

Prestige was exactly what an explorer, or even better, a hero could bring to his nation. Speaking of that, it would be a shame to ignore one of the most important explorers in the history of exploration: James Cook.

Cook can be considered a member of the first generation of British explorers. He joined the Royal Navy in 1757 and worked in collaboration with the Royal Society. In his long career he surveyed, collected data and explored from the Canadian coast to the Antarctic including Australia and his numerous islands charting also the Society Islands, the New Hebrides, New Caledonia and the Hawaiian Islands. Over there, he found his death after a violent encounter with some natives.

It can be said that his contribution to the exploration did not left further major discoveries to his successors. Among the various statues and monuments erected in his honor all over the world today, there is one in his hometown in Chalfont St Giles, in England where on the plinth the visitors can read a sentence that celebrates the explorer: 'To the Memory of Captain James Cook The ablest and most renowned Navigator this or any country hath produced'.[21]

---

[21] Andrew C. F. David, *'Cook, James (1728 –1779)'*, Oxford Dictionary of National Biography, (Oxford University Press, 2004); online edn, Jan 2008

This is something that ought to be taken in consideration: Cook is still recognized by institutions and the general public as an icon, a role model and someone to whom dedicate a statue. But let's take a step back and see what his contemporaries thought about him. It is known that by the end of his career, Cook was generally considered as one of the best navigators of his times and his journals were read by all Europeans.[22] Cook himself knew that his expeditions fully answered the expectations of his superiors receiving also greetings from the King in some occasions.[23] Furthermore, Cook was so esteemed that during the American war of independence, Benjamin Franklin wrote a letter to the captains of the navy that in case of encounter with Captain Cook, he <<and his people [should have been treated] with all civility and kindness, [...] as common friends to mankind>> allowing them to return to England without being plundered and attacked.[24]

---

[22] Forster, Georg. *A voyage round the world, in His Britannic Majesty's sloop, Resolution, commanded by Capt. James Cook, during the years 1772, 3, 4, and 5. By George Forster, F.R.S. Member of the Royal Academy of Madrid, and of the Society for promoting Natural Knowledge at Berlin. In two volumes.* Vol. 1, printed for B. White, in Fleet-Street; J. Robson, in Bond-Street; and P. Elmsly, in the Strand, M DCCLXXVII. (1777)

[23] George Young, *The life and voyages of Captain James Cook: drawn up from his Journals, and otherauthentic documents and comprising much original information* (London, Whittaker, Treacher, 1836), p. 111

[24]Benjamin Franklin, *The works of Benjamin Franklin*, (Tappan Whittemore, and Mason, 1837), pp. 123-24

But he was not respected only by Europeans and Americans, also indigenous people in Polynesia to whom he got in touch used to pay some respect and good consideration to him. In fact, according to Obeyesekere, Cook's qualities such as sympathy, friendliness, firmness and his eager interests in native societies in Polynesia and the way in which their people organized their lives led Cook to have a respectful behavior towards indigenous. For these reasons, he was recognized a new type of explorer who was according to Beaglehole, a man with real feeling for human rights and decencies.[25]

Despite that, it is known that Cook was killed by native Hawaiians and it could be interesting to analyze what this tells us about the construction the heroic image of him.

Firstly, it ought to be considered that there is broad academic debate about this issue. According to Sahlins, the natives interpreted Cook to be a manifestation of the God Lono because of a coincidence: Cook physical resemblance and behavior combined with the fact that he arrived during the Makahiki season, a festival in which Hawaiians celebrated Lono's annual return. In the Hawaiians tradition, The God Lono was supposed to arrive and then leave without coming back for at least one year.

---

[25] Obeyesekere, Gananath. *The Apotheosis of Captain Cook: European Mythmaking In the Pacific.* E-book, Princeton, N.J.: (Princeton University Press, 1997) p.3-4

When Cook sailed, his crew had some damages to the ship and they had to get back to the island in order to make some reparation. This triggered the natives who interpreted the event as a bad omen. As a consequence, Cook lost his appeal on natives and eventually they killed him. In contrast, Obeyesekere has formulated a different theory. He suggested that Cook was celebrated as a chief rather than a God and that the European anthropologists and historians during the years, because of their imperialistic and eurocentric bias, misinterpreted the event. Basically, the Europeans created the myth of Cook, the great explorer and hero, the symbol of western civilization, mistaken for the irrational natives as a God.[26]

If Obeyesekere was actually right in his interpretation, this theory would demonstrate that European had an incredible high consideration of Cook without even realizing it.

The debate is still on today and despite it is really interesting, there is another consideration worthy to be made. As it happened for Livingstone, the way in which the explorer die is a strong element that contribute to his depiction as hero. A tragic death became the coronation of a life devoted to a mission and a dream that can be the scientific eager of knowledge or the greatness of the Empire, or both.

---

[26] Robert Borofsky, Cook, Lono, Obeyesekere and Sahlins, *Current Antropology*, Vol. 38, No. 2 (April 1997) p. 252

In this case, Cook was a scientist and focused more on collection of knowledge and information rather than conquest.

However, he has been celebrated and used by the British Empire as a symbol and example in their imperial propaganda.

The best place where it is possible to find the demonstration of the high value that the British Empire gave to one of its main explorer and hero is the subsequently period in which Cook lived, the Victorian Age. As aforementioned, the nineteenth century was crucial for the history of exploration.

Expeditions in Africa, Australia, in the North Pole and in Antarctica involved a significant number of people born and raised in a particular environment in which exploration and being and explorer had a new meaning. In fact, even if explorer as Cook were esteemed during the eighteenth century, it was not until the following century that they became a distinct species.

According to Adriana Craciun, they started to have their own private clubs and professional organizations. It is not a coincidence that the all the clubs such as the Raleigh Club, the Royal Geographical Club, the Hakluyt Society and so on were founded during the nineteenth century. This was possible thanks to the Victorians rediscovery of Elizabethan exploration, a period in which Cook surely dominate as best example.

This was a nationalist project in which Victorians aimed to create their own precursors, someone to take as inspiration for new adventures.[27]

Before the Victorian Age, the explorer was considered negatively as a spy while someone as Cook would have been regarded more as a scientist rather than an explorer. The explorer before was a definition quite far from the heroic one that in contemporary times we inherited from the Victorians. It was only thanks to them and to men as Cook that the concept of explorer could be reformulated and reinvented.[28] This tells us a lot about the importance that the history of exploration and the exploration itself had for the British, and more widely, for the European society during that time.

The myth of the explorer as hero became in that period inevitably linked to the way in which the new generations were educated. Children's books narrating hero's feats such as the voyages of Cook proliferated at that time. He became a moral exemplar and his adventures narrated for children included ethical and religious social codes to learn from.

---

[27] Adriana Craciun, *'What is an Explorer?'* in Martin Thomas (ed.), Expedition into Empire: Exploratory Journeys and the Making of the Modern World, (New York: Routledge, 2015) p. 26

[28] Adriana Carciun, *"What is an explorer?"* p. 27

For instance, R.M Ballantyne published in 1869 *The Cannibal Island* which incipit about Cook states:

<< Captain Cook was a true hero. His name is known throughout the whole world wherever books are read. He was born in the lowest condition of life, and raised himself to the highest point of fame. He was a self-taught man too. [...] It is the glory of England that many of her greatest men have risen from the ranks of those sons of toil who earn their daily bread in the sweat of their brow. Among all who have thus risen, few stand so high as Captain Cook.>>[29]

At this point it should be clear what meant to be an explorer and why this role was perceived as heroic. It has been shown also that this figure changed over time and that the reinterpretation for this role during the Victorian Age influenced the perception of Cook. In other words, what has been written on Cook after his death, made him the legendary figure that is still considered today.

If Cook is regarded as a member of the first generation of the hero-explorer, while Livingstone and Stanley are members of the second one, Robert Falcon Scott belongs to the third one.

---

[29] R.M. Ballantyne, *The cannibal island, Captain's Cook Adventures in the south Seas* (London, 1869), chapter 1

Scott joined the Royal Navy in the late Victorian Age, in 1880. At this point in history, the culture of exploration was at its culmination. The interior of Africa was no longer a mystery for European, while Antarctica still had a lot to be discovered. Therefore, Scott focused on that part of the world and sailed for his first expedition in 1901. The purpose, once again, was devoted to science. Collect data and gather information, exploring the eastern extremity of the ice barrier discovered by Sir James Clark Ross in 1841 and to search for the land believed by Ross to lie to its east. The voyage ended only three years later.

Considering his lack of previous experience, Scott managed to complete his scientific program spending two consecutive winters in a high latitude in Antarctica, something that had never been done before. This expedition already consecrated Scott as a national hero: he was promoted to captain; the Royal Geographical Society gave him a golden medal as recognition and he was awarded by many countries.[30]

However, what made Scott a legendary figure was the expedition in which he tried to reach the center of the south pole.

---

[30] H.G.R. King, 'Scott Robert Falcon [Known as Scott of the Antarctic] 1868-1912', Oxford dictionary of national biography, Oxford University Press 2004; online edn, Sept 2004

In 1910 he sailed for his last adventure but at the same time, he was aware to be in a rush against his rival Roald Amundsen who was trying to achieve the same objective.

Scott's crew was composed of four more men and none of them managed to survive. Physically deteriorating in the face of low temperatures and adverse wind, with a shortage of food and fuel, the entire group died in his homeward.

The only source testifying what happened it is Scott's diary, a heartbreaking evidence of the journey that took their life.[31]

Despite that, every hero carries the burden of some critics and Scott is no exception. Roland Huntford a famous historian, questioned Scott's heroism arguing that he put at risk his men life on behalf of his own reputation while Amundsen was actually more organized, efficient and prudent. Even though, Scott is still considered a hero because the tragic expedition restored feelings of pride and patriotism among England.

---

[31] H.G.R. King, 'Scott Robert Falcon [Known as Scott of the Antarctic] 1868-1912', Oxford dictionary of national biography, Oxford University Press 2004; online edn, Sept 2004

As a demonstration, a Chicago's newspaper reported that his funeral was comparable only to the Kings Edward's funeral, many members of the British Government, diplomats and high ranks officials from the navy and the military attended the event in order to pay tribute to him.[32]

Furthermore, the Manchester Guardian celebrated the dead explorers as true men.[33]

This is the last characteristic that will be analyzed in this research.

Three out of four among the heroes took into account actually died during their mission and as it has been shown for Livingstone and Cook, that this was essential for the construction of the hero. This happened also with Scott but there is something more. In Scott's case, emerge the characteristic of masculinity consciously constructed by these heroes and by those who celebrated them. In Antarctica, more than everywhere else, the challenge was man versus nature. In this case, the explorer become a hero by surviving in a hostile environment, risking his life not only against violent natives and unknown territories but also in places in which human life is totally impossible.

---

[32] An, British Nation pays tribute to Captain Scott and his four comrades *The day book. [volume]* (Chicago, Ill.), 14 Feb. 1913

[33] Carolyn Strange, Reconsidering the "Tragic" Scott Expedition: Cheerful Masculine Home-making Antarctica, 1910-1913, *Journal of social history,* Vol. 46 No. 1 (Fall 2012) p. 67

Into their dairies, Antarctic traveler, collected practical information which were even more interesting than those gathered into the interior of Africa because the environment was far more dangerous and inhospitable and as a result, the story produced and published after Scott's death was really fascinating for the general public. [34]

Scott's story is useful for underlying a pattern that is possible to be observed in Livingstone and Cook as well: an explorer who died during his expedition is more likely to be celebrated as hero, even if some criticism may persist, than someone as Stanley who survived to his travels. Additionally, Scott's feats highlight the component of masculinity in the heroic figure of the hero. This does not mean that this masculinity is not questionable at all, as Carolyn Strange demonstrated in his article *Reconsidering the Tragic Scott Expedition: Cheerful Masculine Home-making Antarctica*. On the contrary, this is the proof that the construction of the explorer as hero is something that involve an entire society that decide how perceive this figure, how to celebrate it and how to narrate it to the following generations.

It ought to be said that all these men were different from each other. Some of them were more interested in science rather than success or conquest and this is what makes all of them unique in their own way.

---

[34] Carolyn Strange, Reconsidering the "Tragic" Scott Expedition, pp.67-69

However, they all have in common the fact that they were celebrated by their nation, taken as role models and sometimes also as storytellers useful for entertain or even projecting values into the young generation, fostering a wide spread culture of exploration that is still affecting the contemporary perception of the theme in our society as the BBC poll unintentionally proved.

Bibliography

Boehmer Elleke (eds), *Empire Writing, an anthology of colonial literature 1870-1918*, New York, 1998

Driver Felix, *Geography Militant: Cultures of exploration and empire*, Oxford: Blackwell, 2001

Craciun Adriana, 'What is an Explorer?' in Martin Thomas (ed.), *Expedition into Empire: Exploratory Journeys and the Making of the Modern World*, New York: Routledge, 2015

Ballantyne R.M., *The cannibal island, Captain's Cook Adventures in the south Seas*, London, 1869

Forster, Georg. *A voyage round the world, in His Britannic Majesty's sloop, Resolution, commanded by Capt. James Cook, during the years 1772, 3, 4, and 5. By George Forster, F.R.S. Member of the Royal Academy of Madrid, and of the Society for promoting Natural Knowledge at Berlin. In two volumes. Vol. 1, printed for B. White, in Fleet-Street; J. Robson, in Bond-Street; and P. Elmsly, in the Strand*, M DCCLXXVII. [1777]. Eighteenth Century Collections Online, Available at https://link-gale-com.ucd.idm.oclc.org/apps/doc/CB0130087550/ECCO?u=dublin&sid=ECCO&xid=1fde4f50. [Accessed 11 Apr. 2020]

Greely A. W., *Explorers and Travelers* (New York, 1894) Available at https://www.gutenberg.org/files/36069/36069-h/36069-h.htm#page349 [Accessed 14 Apr. 2020]

Franklin Benjamin, *The works of Benjamin Franklin*, Tappan Whittemore, and Mason, 1837

Young George, *The life and voyages of Captain James Cook: drawn up from his Journals, and otherauthentic documents and comprising much original information*, London, Whittaker, Treacher, 1836

Obeyesekere, Gananath. *The Apotheosis of Captain Cook: European Mythmaking In the Pacific*. E-book, Princeton, N.J.: Princeton University Press, 1997, p.3-4 Available at https://hdl-handle-net.ucd.idm.oclc.org/2027/heb.03590. [Accessed 11 Apr 2020]

Robert Borofsky, Cook, Lono, Obeyesekere and Sahlins, *Current Antropology*, Vol. 38, No. 2 April 1997

Carolyn Strange, Reconsidering the "Tragic" Scott Expedition: Cheerful Masculine Home-making Antarctica, 1910-1913, *Journal of social history*, Vol. 46 No. 1 Fall 2012

Matt Wells, 'The 100 greatest Britons: lots of pop, not so much circumstances', *The Guardian,* 22 August 2002, available at https://www.theguardian.com/media/2002/aug/22/britishidentityandsociety.television, [Accessed 15/04/2020]

An, John's Livingstone Testimony, *The New York Herald*, September 09 1872

An, The Slave Trade in Eastern Africa, *The New York Herald*, August 13 1872

An, Livingstone, *The New York Herald*, July 15 1872

An, Doctor Livingstone, Grand Triumph for England and America, *The New York Herald*, July 26 1872

An, Livingstone Herald Special from Central Africa, *The New York Herald*, July 2 1872

An, British Nation pays tribute to Captain Scott and his four comrades, *The day book. [volume]* , Chicago, Ill., 14 Feb. 1913

David Andrew C. F., 'Cook, James (1728 –1779)', Oxford Dictionary of National Biography, Oxford University Press, 2004; online edn, Jan 2008

King H.G.R., 'Scott Robert Falcon [Known as Scott of the Antarctic] 1868-1912', Oxford dictionary of national biography, Oxford University Press 2004; online edn, Sept 2004

Moody Blues, Dr. Livingstone, I Presume, Youtube video available at https://www.youtube.com/watch?v=TPkc6VOsqZ0, [Accessed 12/04/2020]

TRAVELING IN THE SERVICE OF SCIENCE

Introduction

This essay will be discussing the importance of science and its role in exploration. Starting from the role of the Royal Geographical Society I will argue that explorers at that time used science not only for collect information and gathering knowledge but also for helping the European empires in expending their control over the world.

Eventually I will analyze a primary source and a representative example of explorer such as Mungo Park and what kind of role he had as one of the firsts explorers in the inner Africa.

In his travel it will be possible to find a lot of elements discussed in the aforementioned sections of the essay.

1 - The importance of the Royal Geographical Society

The Royal Geographical Society (RGS) was founded in 1830 in England and played a pivotal role in mid-Victorian age. The RGS redefined science's frontiers drawing a connection between different explorers unified by their aim to pursuit knowledge. James Cook, George Forster, Alexander Von Humboldt and Charles Darwin were some of the famous explorers who contributed to the RGS' mission to collect information and data.[35]

For the first time in history, a single hub for explorers was established in order to create a centralized, systematic and scientific approach to geography. The RGS was born because the Royal Society which presidency was held by Joseph Bank, was falling apart. During the nineteenth century, a lot of new scientific societies more specialist were founded. Each of these societies had the purpose to focus on their particular field of studies and this caused a fragmentation and proliferation of new societies and associations.[36]

Furthermore, the RGS gradually developed a new approach to exploration that Susan Cannon called *Humboldtian*.[37]

---

[35] Felix Driver, 'The Royal Geographical Society and the Empire of Science', ch. 2 of *Geography Militant: Cultures of exploration and empire* (Oxford: Blackwell, 2001), pp. 24-26
[36] Felix Driver, 'The Royal Geographical Society and the Empire of Science', ch. 2 of *Geography Militant: Cultures of exploration and empire* (Oxford: Blackwell, 2001), pp. 29-33
[37] Felix Driver, 'The Royal Geographical Society and the Empire of Science', ch. 2 of

According to her, from that moment science started to be studied on field while scientists were becoming more and more involved in travels and explorations, carrying with them instruments like chronometers, sextans, horizons and barometers in order to measure and map everything.

However, scientific discoveries were not the only interests of those travelers. In fact, Felix Driver pointed out that the Empire was essential for explorers because most of their studies depended on the navy or the military that could provide logistic support which turned out to be really useful in some voyages. On the other hand, those scientists and geographers successfully assisted the Empire by collecting and providing all kind of useful information that have been used in order to trade or to control some territories. Nevertheless, for the RGS its purpose of gathering knowledge would have been essential not only for the empire but for all the humanity and would have fostered global commerce. [38]

As a consequence, the military and government resulted to be intertwined with the RGS. As a matter of fact, the firsts two presidents of the RGS were colonial ministers and many of their successors were career diplomats. In addition, army and navy officers constituted around one fifth of the founding members.

Geography Militant: Cultures of exploration and empire (Oxford: Blackwell, 2001), pp. 34-35

[38] Felix Driver, 'The Royal Geographical Society and the Empire of Science', ch. 2 of *Geography Militant: Cultures of exploration and empire* (Oxford: Blackwell, 2001), pp. 37-42

Between 1850 and 1870 Roderick Murchison has been many times president of the RGS and in his opinion, science was an instrument for imperial expansionism. In his position he was able to organize a scientific team made up by a geographer, a botanist a meteorologist and a geologist which would have followed a military expedition in Abyssinia in 1867. The main purpose of this expedition was to punish the Abyssinian king who had kidnapped a British citizen.

In that occasion Murchison stated:

<<When has Europe marched a scientifically organized army into an unknown intertropical region, and urged it forward as we have done, for hundreds of miles over chain after chain of Alps amidst the grandest scenery? And all to punish a dark king... This truly is a fine moral lesson which we have read to the world; and...; in addition, we reap good scientific data.>>[39]

This is probably the most evident example of how close together were the army, the governmental institution and the RGS.

Despite that, is quite important to underline the fact that the RGS had its own independence from the government and it could not act as a single lobby or a homogeneous entity and it could seldom speak with the same voice.

---

[39] Felix Driver, 'The Royal Geographical Society and the Empire of Science', ch. 2 of *Geography Militant: Cultures of exploration and empire* (Oxford: Blackwell, 2001), p. 44

Some members had different opinions and an interesting example of that is Francis Galton and Thomas Hodgkin quarrel regarding the supposedly genetically inferior black people. While Galton believed in white superiority, Hodgkin advocated for the unity and equality of human race.⁴⁰

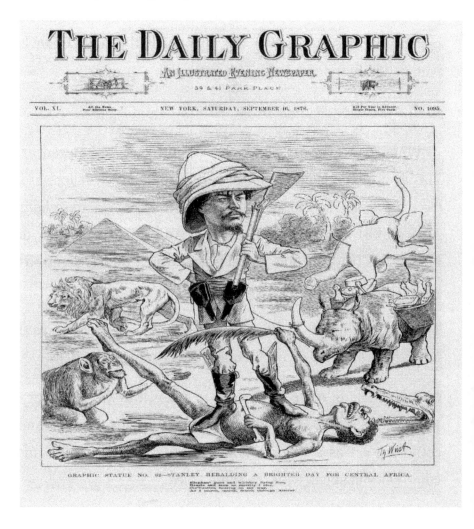

---

[40] Felix Driver, 'The Royal Geographical Society and the Empire of Science', ch. 2 of *Geography Militant: Cultures of exploration and empire* (Oxford: Blackwell, 2001), pp.45-46

2 – Not only a British effort

The British explorers were not the only ones to explore Africa. In the effort participated also French, Belgian and Germans explorers. Each of them had a different approach to the exploration.

While the Belgians used to give more importance to team and they named the expedition after chronological numbers, the Germans used to evaluate more the explorer who led the travel and because of that, they named the expedition after his name.

In 1876, Leopold I king of Belgium founded the International African Association which purpose was to foster explorations in the interior of Africa. This association had an immense importance in supporting adventurers and scientists during their travels, including the Belgian explorer, Jerome Becker.

Becker wrote *"La vie en Afrique"* in which he recorded practical advice for whom found itself in the need of travelling in harsh condition in Africa. According to Becker and his guide, Europeans who were exploring Africa had to forget about the division of labor because in those difficult conditions every member of the team had to be able to do every kind of job.

In addition, in his vedemecum, Becker gives some advice regarding the equipment, supplies food and number of men that an aspiring explorer is supposed to bring with him.[41]

Moreover, Becker underlines the necessity of being always ready to prevent accidents and political conditions. In fact, Africa was far from be completely ungoverned. There were multiple authorities, rulers, tribes and kings upon depended the possibility to get safely inside some territories.

Despite that, European nations did not consider African ruler as their equals and this meant that international diplomacy have not been used in Africa, except for addressing other European powers.

Becker's vademecum was not the only guide for travelers.

Also, Francis Galton wrote a manual about what was useful to bring during an expedition. This is the demonstration of a new phase in geographical exploration. In this phase travels are common, well organized, completely fostered by governments and to their contribution participated explorers and scientists from all over Europe in a sort of common effort toward knowledge and territorial conquest.

---

[41] Johannes Fabian, *Living and Dying*, ch.2 of *Out of Our Minds: Reason and Madness in the Exploration of Central Africa* (Berkeley: University of California Press, 2000), pp.52-60

In this context the explorer became a figure depicted as a hero.

Their gestures were narrated in books wrote often by themselves and published in journals and newspapers. A remarkable example of this is the famous adventure of Henry Morton Stanley who was traveling in the interior of Africa and found David Livingstone who was thought lost by the RGS.

Additionally, the International African Association gave the task of founding scientific station to its explorers. On one hand, this could help gathering information, collecting data on fauna and flora and helping geographers to map the territory.

On the other hand, the scientific stations turned out to be necessary for controlling militarily entire regions of Africa.

Gradually, exploration reached the end of the road and tuned into colonization.[42]

---

[42] Johannes Fabian, *Living and Dying*, ch.2 of *Out of Our Minds: Reason and Madness in the Exploration of Central Africa* (Berkeley: University of California Press, 2000), pp.70 -77

3 – A primary source and an outstanding example

In order to comprehend the relevance of science and geography in the exploration of inner Africa, it would be useful to analyze the role held by Mungo Park, a Scottish explorer who traveled in India and west Africa. His first journey was published in 1799 in a successful book called "Travels" which brought him great celebrity.[43]

In 1795, the African Society hired Park for a Herculean task: finding the source of the Niger river. Additionally, in his voyage, Park wanted also to analyze Western Africa flora. The African Society had already hired the Major Houghton who unfortunately died during his journey in the inland.

Starting his journey in Pisania, the modern Gambia, Park spent his firsts days learning Mandingo tongue which would have been useful to communicate during his long travel and in order to get an extensive knowledge of the country. [44]

In his journal, Park describes accurately some native groups. Europeans knew that if they wanted to have the chance to trade with the inner tribes, they had to know them. Those kinds of information were really important not only for the African Association but also for governments.

---

[43] Christopher Fyfe, 'Park, Mungo (1771 – 1806), Oxford Dictionary of National Biography, Oxford University Press 2004

[44] Mungo Park, vol. 1 of *Travels in the Interior of Africa* (1799), chapter I

Additionally, in the European conscience was growing an idea of superiority in those years. In fact, Park stated:

<<How greatly is to be wished that the minds of a people so determined and faithful could be softened and civilized by the mild and benevolent spirit of Christianity>>[45]

With this statement is easy to understand that another purpose was to civilize and convert those African tribes. Even though Park himself recognize that not all tribes were as much as backward as European could have thought. As a matter of fact, the Mandingoes had a really complex organization and power structure. They had a magistrate called *Alkaid* whose office was hereditary and whose business was to administrate justice. They had not written language so the related on ancient customs; but since the Islamic religion spread so far, they used the Koran and the Sharia. The Mandingoes also had lawyers whose goal was to solve controversies. A master for example, could not kill or sell a slave without a trial. This was an important law because according to Park three fourth of the population was slave.

The slavery issue was discussed in England in those years and many people advocated for its abolishment in Europe. However, in that part of Africa, slaves made up the main part of trade together with iron and other minerals. Still was a profitable business.

---

[45] Mungo Park, vol. 1 of *Travels in the Interior of Africa* (1799), chapter II

When Park left Pisania he travelled with a team formed by five men more. Two of them were interpreters and cultural brokers that helped him during the journey. Johnson spoke both English and Mandingo tongue because he had been a slave in Jamaiaca and he gained his freedom. Whilst Demba was a slave who spoke Mandingo tongue and Serawolli tongue. The crucial role of the broker has been already mentioned several times during this module.

Park encountered some kings and rulers during his long voyage. In every kingdom he entered, he had to bestow something to the King of that territory. That was the only way to pass safely through the territory. This is interesting because makes clear what kind of difficulties every explorer found during their expeditions which is underlined also by Becker as it has been already mentioned. [46]

It was only in the Ludamar territory that Park was detained by the Moors and his king Ali. In this occasion the king did not want him to leave because he wanted to show him to his wife. I would suggest that this is quite curious. For the first time in their life, most of the Moors met a white man and they had a lot of questions for him and showed great interest in many aspects

---

[46] Mungo Park, vol. 1 of *Travels in the Interior of Africa* (1799), chapter III - VIII

of him. I would say that this was for them an occasion to *explore the explorer*.⁴⁷

After having been in captivity for some month with little food and water, Park finally managed to escape and eventually accomplish his mission of mapping a great portion of the Niger River coming back from Africa safe and sound.

Unfortunately, he died during his second travel there but he definitively set a standard for all the explorer that followed him and that is why he is so fundamental.⁴⁸

Conclusion

In conclusion I would underline the importance of militant science as instrument of exploration and subsequently of military conquest. This is a link between these elements and even if a great number of explorers had noble intention and merely scientific interests, sometimes their efforts have been used as practical information by armies and governments with or without their consensus. I think that this research enlightens both aspects of scientific exploration.

---

⁴⁷ Mungo Park, vol. 1 of *Travels in the Interior of Africa* (1799), chapter IX - XII

⁴⁸ Christopher Fyfe, 'Park, Mungo (1771 – 1806), Oxford Dictionary of National Biography, Oxford University Press 2004

Bibliography

Johannes Fabian, *Living and Dying, ch.2 of Out of Our Minds: Reason and Madness in the Exploration of Central Africa* (Berkeley: University of California Press, 2000), pp.52-77

Felix Driver, 'The Royal Geographical Society and the Empire of Science', ch. 2 of
Geography Militant: Cultures of exploration and empire (Oxford: Blackwell, 2001), pp. 24- 48.

Christopher Fyfe, 'Park, Mungo (1771 – 1806) , Oxford Dictionary of National Biography, Oxford University Press 2004

Mungo Park, vol. 1 of *Travels in the Interior of Africa* (1799)
http://www.gutenberg.org/files/5266/5266-h/5266-h.htm

THE NON-ECONOMIC REASONS FOR MAKING LIBYA A COLONY

There were many reasons for European powers to pursuit the *grandeur* during the imperialistic and colonialist period. One of these was certainly the objective to expand their economic and trade influence over some territories in order to gain advantages. During the 19th and 20th century, European powers struggled to keep a peaceful and stable situation in Europe while they were expanding all over the world which was seen as a necessary policy in order to keep the other European competitors down. In that period Italy gained its independence and unified in a single country and its perception was to be a weak country surrounded by superpowers with their own economic and political influence even in the Mediterranean which was supposed to be an Italian lake in Italy's propaganda.

The main aim of this essay is to analyze to what extent trade and economy influenced Italian's decision to colonized Libya in the early years of the 20th century and whether the attempt of *"penetration pacifique"* succeeded or not. I will argue that rather than seeking for a natural end market for their goods, Italians were looking more for a *place in the sun,* in order to bridge the gap with the other European powers, gain prestige and expand their territory.

The economic importance of Libya was questionable. In the past, the trans-Saharan trade system had flourished mostly because of black slaves trade which was abolished in Tripoli ruled by the Turkish in 1857 causing a great lost in the economic Libyan income. Additionally, French and British control over modern Senegal and Niger diverted the traditional trade route with caravans across the desert to reach Libya to new trans-Atlantic destinations. In a short time at the end of the nineteenth century, trading with Europeans became cheaper for western African merchants. By the time in which Italy invaded Libya in 1911, the trade importance of Libyan routes had already reached its decline. [49] However, also in the Mediterranean Sea Europeans merchants imposed their supremacy with Africa. By the 1902 Great Britain was Libya's principal agricultural export market followed by the Ottoman Empire, France, Tunisia, Italy and Germany.[50] But trade is not the only wealth of a country. European explorers had been overestimating the agricultural potentiality of Libya since their first travels in its provinces.

---

[49] John Wright, *A history of Libya*, (London 2012), p.95

[50] Lisa Anderson, *The state and social transformation in Tunisia and Libya 1830-1980*, (Princeton university press, 1986), p.111

In fact, according to Dutch consul opinion in 1857, Tripoli could have become the first town of North Africa and the country the granary of Malta, Marseilles and Paris.[51]

Both explorers Gerhard Rohlfs and Manfredo Camperio agreed that Libya could have been a farming colony suitable for Italians and that the government could have hugely boosted the trades in the region. It was thanks to Camperio that Libya started to be considered the <<Promised Land>> or an <<Eldorado>> with <<fecund land>> and <<fairy gardens>>.[52]

Italian policy makers had always been worried about the danger of being kept as a hostage in the Mediterranean Sea. Already in 1863 on the *Opinione* of Turin a journalist stated that <<If Egypt and with it the Suez Canal falls to the British, if Tunis falls to the French, and if Austria expands from Dalmatia into Albania, we will soon find ourselves without a breathing space in the dead centre of the Mediterranean>>.[53]

Actually, Italians were more interested in Tunisia rather than Libya. In the former in the 1880s there was already a community of 25.000 Italians.

---

[51] Wright, *A history of Libya*, p.97
[52] Angelo Del Boca, *Gli italiani in Libia, Tripoli bel soul d'amore 1860-1922*, (Milano, 2015), pp. 8-10
[53] Wright, *A history of Libya*, p. 103

Eventually, France succeeded in taking control over that territory before Italy thanks to the Berlin Conference and as a consequence, Italian imperial policy focused on the only north African territory still available.

However, French and Italian rivalry did not ended there because the partition of north African's borders have been disputed during all colonial period.[54]

Another important Italian politician, Francesco Crispi, thought that the geopolitical situation of his country was strategically and economically strangled by other Europeans power in the 1880s. According to Crispi, the Austro-Hungarian empire, the British and the French, with their presence in the Mediterranean, could threaten Italian industries and trade.[55] As a consequence, the only solution to this problem was to find a colony as a space of safety and strategic defence against European powers in the Mediterranean. On account of that, in 1883 it was founded in Naples the Comitato Italiano per la Tripolitania which aim is to foster Italian institution in that region.[56] This was the beginning of the so-called attempt of *"penetration pacifique"*.

During the 80s the Italian diplomacy focused on firming the Triple Alliance which considered Italian influence over Libya.

---

[54] Wright, *A history of Libya*, p. 108
[55] Del Boca, *Gli italiani*, p.12
[56] Del Boca, *Gli italiani*, pp. 13-14

Furthermore, Corti, the ambassador in London agreed with the British prime minister on the maintaining of the *status quo* over the Mediterranean Sea while an agreement with the Spanish government prevented Spain to cooperate with France against Italian will.[57]

In the following 90s and early years of the 20th century, Italian and French ambassadors managed to reach an agreement to their reciprocal influence over north Africa while in 1909 also the Russian empire recognized Italian interests over Libya. This was an intensive and necessary work for Italian diplomacy since it would have been impossible to impose their interests without acknowledge the other European powers.

In the meantime, the increasing interest of Italy into Libya started to annoy the Ottoman mutaserrif of Benghazi who in 1901 tried to stop the Italian attempt to open a postal office in the city. Only the threaten of two battleships in front of Benghazi allowed the Italians to establish their postal office.[58] This was the demonstration that in case of peaceful economic penetration in the region, the Italian government was ready to intervene militarily.

---

[57] Del Boca, *Gli italiani*, p.17
[58] Del Boca, *Gli italiani*, p.24

in 1905, the Banco di Roma, linked both to the Italian government and to the Vatican, started to peacefully and economically penetrate the country by financing new business, controlling interests in shipping lines between ports that exported from Libya esparto grass, cereals, ostrich feathers, ivory, wool and sponges.[59] The main aim was to imitate what France had done for Tunisia and Britain for Egypt.[60] Additionally, in 1907, the Banco established some branches in Tripoli, Benghazi, Zlitin, Khum and Misratah. In two of these cities the Ottoman Empire did not even authorized the operation. Moreover, Ottomans opposed the Italian economic expansion trying to prevent the local population from establishing a mutual beneficial relationship. Despite Italian efforts, in the 1911 just before the invasion, Libya did not have an extensive trade relation with Europe and few notables of shore cities were really entangled with Italian finance.[61] However, the Banco realized lots of industries and commercial activities in the country. By 1911 the Banco had invested more than 5 million of dollars in Libya.

---

[59] Wright, *A history of Libya*, pp. 105-106
[60] Nicola Labanca, *La Guerra italiana per la Libia 1911-1931*, (Bologna 2012), p.35
[61] Anderson, *The State and Social transformation*, pp.111-113

This behaviour and the over activity of the Banco, lead the Ottomans to be suspicious of the Italian bank and as matter of facts they let grow in the native population hostility towards Italians while explicitly obstructing the Banco's effort to penetrate in the region.[62] Therefore, the two side of the Mediterranean Sea did not result to be so entangled as the Italians wanted at that time.

In 1911 it was clear that the peaceful penetration had failed. The reasons were mainly two. On one hand Turkey hostility had occasionally impeded Italian initiatives and on the other hand, Ottoman officials helped also other European countries to develop economic initiative in their territory.[63] At this point, the Italian nationalist party tried to push the public opinion and the government into the war. Their main argument was that the Italian conquest of Libya would have granted Italian strategic security in the Mediterranean and it would have paved the way for Italian control over the East Mediterranean.[64] This was without any doubt a necessary hyperbole to persuade Italian society to occupy that region.

However, the nationalist party was not the only one to stand for a war.

---

[62] Del Boca, *Gli Italiani*, pp.37-44
[63] Renato Mori, 'La penetrazione pacifica italiana in Libia dal 1907 al 1911 e il Banco di Roma', *Rivista di Studi Politici Internazionali*, Vol. 24, No. 1 (January – March 1957), pp. 102-118, p. 115
[64] Angelo Del Boca, *Gli italiani in Libia*, p.53

Generally speaking, with the exception of some Socialists, the press agreed with the nationalist's argumentation. In their view, Libya was a rich country that would have granted to Italy wealth and security if taken.[65] The great majority of the policy makers thought that Libya was really a promised land with great potentiality especially as a farming land. This illusion fed many dreams among Italian colonialists and it was used as a justification for the military occupation of the region,[66] but it was far from reality.

A more plausible explanation concerns other reasons.

On top of that, the international context was changing considerably. Between 1910 and 1911 occurred a second Morocco crises that opposed France and Germany. For Italy there was the risk that the European powers could agree on a different future for Libya.[67]

At this point, the Italian government pushed by the public opinion, the press, the nationalists and the international situation, chose to occupy Libya on account of Ottoman obstructionism of Italian economic penetration.

---

[65] Angelo Del Boca, *Gli italiani in Libia*, p.56
[66] Nicola Labanca, *Oltremare*, (Bologna 2012), pp. 108-115
[67] Nicola Labanca, *La guerra italiana*, p.36

The official and diplomatic motivation for going in an open conflict against Ottomans was Italian protection of its economic interests[68] but this was only partially true because in the *ultimatum* that the Italian government sent to the Turkey counterpart there was not a significant *casus belli*. What is more, the Turkish prime minister, in an extreme attempt of pacification, tried to stop Italian occupation of Libya by drastically reconsidering ottoman obstructionism to economic penetration.

However, this attempt turned up to be useless because Italian government had already chose to invade the Ottoman possession as a demonstration of the lack of interest into a merely protectorate which would have limited Italian influence on a simple economic sphere.[69]

This was a significant demonstration that Italy did not merely seek for a limited economic exploitation. On the contrary, the main achievement would have been a complete control over Libya.

In fact, in his colonial history Italy could be considered an exceptional case because its imperialism was not mainly motivated by economic interests. As it has been explained, the Banco di Roma's attempt to gain control over the economic life of Libya slowly and without being noticed, failed.

---

[68] Mori, *La penetrazione pacifica italiana*, p. 102
[69] Del Boca, *Gli italiani in Libia*, p.76

Even though trade and economic reasons were not the main motivation for the colonization, still Italian colonialism had its own economic system able to shape political choices but it has been widely studied by scholars the fact that Italian colonialism was a phenomenon characterised by a really poor economic interest during the first phase of colonial expansion.

Almost all governments were more interested in Italian prestige and reputation at international level rather than expanding the market or ensuring trading routes as the British Empire usually did.[70]

In conclusion, I would suggest that in the case of Italian occupation of Libya the economic interests were not the main reason why the war began. On the contrary, the Italians sought more strategic and geopolitical stability that would have resulted, in their hopes, in international prestige. This unusual colonial experience in which economy and trade occupy a smaller role than national propaganda, international reputation and the necessity to prevent other European powers to expand themselves in a Sea perceived as a national environment, should deserve more attention.

---

[70] Labanca, *La guerra italiana*, pp. 30-32

Bibliography

John Wright, *A history of Libya*, London 2012

Lisa Anderson, *The state and social transformation in Tunisia and Libya 1830-1980*, Princeton university press, 1986

Nicola Labanca, *La Guerra italiana per la Libia 1911-1931*, Bologna 2012

Nicola Labanca, *Oltremare*, Bologna 2012

Angelo Del Boca, *Gli italiani in Libia, Tripoli bel soul d'amore 1860-1922*, Milano, 2015

Renato Mori, 'La penetrazione pacifica italiana in Libia dal 1907 al 1911 e il Banco di Roma', *Rivista di Studi Politici Internazionali*, Vol. 24, No. 1 (January – March 1957), pp. 102-118

THE POSTCOLONIAL "CRISIS" IN THE NATIONAL BUILDING OF ALGERIA, SYRIA AND IRAQ

The contemporary Middle East has faced some troubles in its post-colonial period. As a matter of fact, most of the states that emerged after the colonization have been struggling with a national building effort in which they attempted to create a coherent and united state with its tradition and elements of identity. This elaborate will argue that the colonial period has caused some difficulties for the post-colonial government in building and defining the new nation. This challenge has been taken on in multiple ways, and it will analyze the Algerian Iraqi and Syrian example briefly.

Furthermore, the governments of former colonies have inherited boundaries already drawn on maps, and it will be claimed that this caused even more problems in the national identity building effort by referring to the Sykes-Picot agreement in Syrian and Iraqi context. Despite that, it will be claimed that the national identity of these states does not depend merely on what colonial powers did during their rule.

On the contrary, they have had an independent development which transcends western influence.

Middle East states were created following a European model because organizations of this type were the "compulsory model" for establishing a national unit outside Europe at that time.[71] After the defeat of the ottoman empire during the World War I, British and French took control over the former Ottoman provinces of Basra, Baghdad, Mossoul, Haleb, Sourtiya and Beyrut.

In that moment, most of the future problems involving this part of the world were emerging, and the way the colonial powers dealt with them produced post-colonial government challenges. In fact, it was the European countries controlling those regions that first created centralized administrations, coherent modern legal system and political structures. In some cases, they did that by assembling some old ottoman provinces as in the case of Syria and Iraq while in other cases using boundaries already formed like the French colony in Algeria.

However, this process would have brought to a lack of ethnic homogeneity in the post-colonial period and someone could argue that the reason for internal fight and political instability lay in this operation.[72]

---

[71] Roger Owen, *State, Power and Politics in the Making of the Modern Middle East*, London 1992, p. 3
[72] Roger Owen, State, Power and Politics p. 9

Additionally, the ethnic, sectarian and tribal division was a widespread common practice among colonial officers who attempted a strategy of *'divide et impera'* in order to achieve better control over the subjugated population. This was true especially for the French colony in Syria[73] and in Algeria.[74]

Moreover, another issue was about the European settlers in Middle East colonies. Especially in Algeria, it has been attempted to prompt colonization by moving in those territories a large number of people with the purpose of farming lands. These groups of settlers had their own privileges with political bodies, control over best lands and exclusive trade unions.[75] When the colonial period ended, the new governments of the former colonies had to deal with this part of the population. Eventually, in Algeria, they were expelled in order to regain control over the territory and its economy.

The majority of Middle Eastern states after the independence had to face a destitute situation and the aim of catching up with the west led them to develop state apparatus and bureaucracy in order to fully control the newborn state and fostering programs of economic development and social welfare.

---

[73] Roger Owen, *State, Power and Politics*, p. 13

[74] Marek Čejka, *Divide et Impera?: Western engagement in the Middle East*, Central European journal of International and Security Studies, Vol 6, Issue 4(2012) pp.200-219

[75] Roger Owen, *State, Power and Politics*, p. 13

This caused the creation of an authoritarian government which control over the society was particularly intense. [76]

This happened in Algeria, Syria and Iraq in similar ways.

As aforementioned, the end of colonial control over Middle East territories left the necessity to build a common and strong identity among its inhabitants with a reliable connection to the government and the nation. This was also the case of Algeria. The day after the independence, Algerians were far from being a single community and the nationalists who fought against France started underlying everyday experiences, ancestors, struggles and triumphs. They also promoted a better future and a fast modernization under a freed Algeria controlled by a single party, the FLN (Front de Libération Nationale). The presidents Ben Bella, Bomedienne and Benjedid encouraged the narration of a nation with "Arab and Muslim essence" ignoring the fact that in Algeria also lived the Berbers.

Additionally, Algeria's society had to be organized according to the principle of socialism. These elements allowed the FLN party to establish firm control over the society as an authoritarian State would do and were even included in the Algerian constitution.

---

[76] Roger Owen, *State, Power and Politics*, p.24

As a consequence, during the 60s and the 70s the government tried to foster an Arabisation in the country by implementing the Arabic language in schools and trying to replace French primacy. However, the attempt resulted in a disadvantage for those students who studied only Arabic because they eventually became less relevant in the labour market.

The government led by president Benjedid responded by creating jobs opportunity for who spoke Arabic and implementing Arabic language also in the universities. This contributed to instigating Berber resentment and caused the so-called "Berber Spring" in which thousands of young Berbers protested demanding the official recognition of their language. Some of them even preferred the French language instead of Arabic.

Each of these elements, Arab culture, Muslim religion and socialism, were essential as post-colonial identity. Socialism, for instance, was supposed to fight capitalism which was perceived as one of the main reasons that led French to colonize Algeria. Basically, the need for new markets, new resources lands and cheap labour had been what originated French invasion.

What is more, socialism at that time was considered the best option for stimulating economic growth and in the meantime did not necessarily clash with Islam which was something that all Algerians shared.[77]

Additionally, socialism had the concept of revolution which constituted the core of the independence war. It is not wrong to say that after the gruelling battle against France, Algeria became a hub for all revolutionary parties in the third world. President Ben Bella also fostered Algerian role as benchmark trying to lead the third world movement. Amilcar Cabral, leader of the nationalist party in Guinea-Bissau, defined Algeria the "Mecca of Revolution". In a short time, during the 50s and the 60s Algerians started supporting other rebels against colonial power all over the world while Cuban and Yugoslavian governments understood that they could only spread their influence in Africa trough their relation with Algeria.[78]

The FLN saw its popularity as an opportunity for promoting a geopolitical movement of non-aligned countries in the broader context of the Cold War.

---

[77] Jonathan N.C. Hill, , Identity and instability in postcolonial Algeria, the Journal of North African Studies pp. 1-12

[78] Jeffrey Byrne, *Mecca of Revolution: Algeria, Decolonization and the Third World Order* (Oxford: Oxford University Press, 2016), p.3

The Third Worldism was perceived both as post-colonial identity and legitimation for decolonization[79] and was essential in the late colonial period and during the early years of independence for the nation-building of the country. The revolution became so fundamental for defining the Algerian nation that eventually when Algeria conquered its freedom and the Third Worldism non-aligned movement started losing importance, the nationalist leaders of the country worried that without the revolution it was impossible to have their nation.[80]

Hence, it seems that the colonial experience drastically influenced the post-colonial state-building in Algeria. Nevertheless, this country was not the only one who struggled to find its identity.

Some countries like Iraq or Syria have been even considered artificial and without national identity because of the colonization and their borders have been questioned multiple times by a large number of scholars, political actors and subversive groups who wanted to rethink the Sykes-Picot agreement.[81]

---

[79] Jeffrey Byrne, *Mecca of Revolution*, p.10
[80] Jeffrey Byrne, *Mecca of Revolution*, p. 298
[81] Sara Pursley, *'Lines Drawn in the Sand'*, Jadaliyya.1

For instance, the contemporary Syrian civil war began in 2011 and some scholars still consider it as the direct consequence of western intervention in the region after the First World War while the Sunni-Shi'ite sectarian conflict in Iraq has been explained blaming the British who created the Iraqi State in 1920.[82] However, Sara Pursley in *Lines drawn on an empty map*, claims that the narrative of artificial state was developed by the British during the 20s in order to prevent Arab self-government. Besides, colonial powers were not the only actors in drawing the borders. In fact, Arab nationalists in Iraq contributed to enlarge their country by pressing the British to keep the city of Mosul away from Turkey's control. Eventually, they even succeeded in that.[83]

In addition, from 1918 to 1929 there were in the Middle East massive revolutionary movement that substantially contributed to the national building formation. In order to understand the population's role in establishing their own borders during the colonial period it should be underlined that the Sykes-Picot agreement was not the only moment in which the boundaries were established. According to Pursley, some scholars argued that the Paris Conference of 1919, the San Remo Agreement of 1920 and the Cairo Agreement in the following year were also moments in which the colonial powers drew lines on middle eastern maps.

---

[82] Liora Lukitz, *The search for national identity* (Frank Cass 1995)
[83] Sara Pursley, *'Lines Drawn in the Sand'*, Jadaliyya part. 2

But the question is why they should have modified their initial intention and the answer is that the local population at that time was generally asking for independence in Syria and Iraq and they influenced somehow European's decision. [84] Consequently, the artificial state narrative is not accurate because, in the process of national identity, the middle eastern population played its role.

Nevertheless, it is undoubtedly true that the question of borders in the Middle East is something in which Europeans played an important role, even if local actors contributed and this inevitably changed how those countries at the end of the colonization perceived their national identity. The point is that a national identity exists in contemporary times and existed even after the First World War when the European powers took control of the region. The thesis of the artificial state can be seen as a western construction in order to explain a more complicated reality in which nation-building formation did not go off without hitches.

It has been often argued that the reasons why ethnosectarianism, authoritarian states and political instability emerged in countries like Iraq or Syria were related to Europeans drawing borders during colonization.

---

[84] Sara Pursley, *'Lines Drawn in the Sand'*, Jadaliyya part. 1

Some scholars and policymakers even proposed to remap the entire region in order to fix the supposed mistakes made by the Sykes-Picot agreement that mixed different ethnicities and sectaries.[85] Although, Daniel Neep claimed that the real problem has been western obsession with maps and its willingness in creating <<[...] **one group, one state**, in which identity and geography are – or, more dangerously, *should be* – essentially indistinguishable.>>[86] In other words, according to Neep, it is impossible to fix political problems with cartographic solutions because the identity of a nation is something that it has to be found within the nation itself, without further interferences.

However, not only western scholars and intellectuals are to blame for this long-standing colonial narrative. Pan-Arab nationalists could have had their role in all this by fostering an historical interpretation in order to gain some advantages from this. One direct effect of colonialism, as it happened for Algeria, is to directly cause anti-colonialism, a reaction in which a population look back at his own traditions and heritage in order to find a cultural model to oppose to the western's one.

---

[85] Daniel Neep, *'FOCUS: THE MIDDLE EAST, HALLUCINATION, AND THE CARTOGRAPHIC IMAGINATION'*
[86] Daniel Neep, *'FOCUS: THE MIDDLE EAST, HALLUCINATION, AND THE CARTOGRAPHIC IMAGINATION'*

In its long development, Arab nationalists in Syria found a standard-bearer in a political party founded in the 1940s by Michel Aflaq and Salah a-Din al-Bitar called *Ba'ath*. Their political purpose is easy to explain but challenging to achieve: to unify all the Arabs country because they all shared a single religion, a shared cultural heritage and a unique language. Complete independence and a true social justice could have been obtained only through a great Arab nation that would have been able to free itself from the constraints imposed by the Conference of Peace held in Versailles in 1919.[87]

It seems clear that the first objective of a pan Arab movement that wants to build a large nation without inner borders is to delegitimize those boundaries in order to make them look like they are unnatural or *"artificial"*. As a consequence, the Ba'athism did not accept Syrian borders and when some military officials got in touch with Gamal Abd el-Nasser, the major pan Arab nationalist leader who had great influence all over the Arab world and was also president of Egypt, they founded the United Arab Republic (UAR) in 1958 which was open in its intention to admit other Arab states.[88]

---

[87] Eugene Rogan, *The Arabs, A history*, New York 2009, p.306
[88] Eugene Rogan, *The Arabs*, p.308

Meanwhile, in Iraq, a movement of Free Officers inspired by Nasser's example in Egypt, organized a *coup d'etat* overthrowing the monarchy in order to create new institutions without links to the old British colonial rulers. Despite that, the members of the new Iraqi government did not agree on what to do in respect of the UAR. Some of them wanted to join the United Arab Republic while others wanted to keep Iraq independent.[89] This is a further demonstration that the colonial period caused not only a wish of independence but also two different nationalisms in the Arab world. A broader pan Arab movement which considered political borders as imposed by westerns and artificial while some other Arab nationalists considered those boundaries as worthy of being defended and not dismissed. In fact, Abd al-Karim Qasim, the leader of the Free Officers in Iraq, was unwilling to merge his country with the UAR because of his strong rivalry with Nasser and Egypt.[90]

Not only Iraq had his national identity. Also in Syria, the members of the Ba'ath party soon discovered that the United Arab Republic was an organism in which Nasser and Egyptian institutions had too much power. As a result, in 1961, the Syrian army organized another putsch for taking control of the country again and excluding Nasser from it. [91]

---

[89] Eugene Rogan, *The Arabs*, pp- 316-317
[90] Eugene Rogan, *The Arabs*, pp. 316-317
[91] Eugene Rogan, *The Arabs*, pp. 320-322

To conclude, the Algerian, Iraqi and Syrian examples show that their long journey for national identity has been drastically influenced by their colonial experience but still they did not wholly depend on that as the historiography generally put it. It would be an oversimplification to assert that Algeria, Syria or Iraq are not actual states because they are too ethnically different or because they struggled in keeping their countries together and in finding a national identity. They certainly had difficulties in doing so and this largely depended on their colonial past that generated more troubles. Until there is someone who believes in a universal nation, there is still a national identity and all these states had people, good or evil is not the point, who fought for keeping the country together.

Bibliography

Eugene Rogan, *The Arabs, A History,* New York 2009

Roger Owen, *State, Power and Politics in the Making of the Modern Middle East*, London 1992

Jeffrey Byrne, *Mecca of Revolution: Algeria, Decolonization and the Third World Order* (Oxford: Oxford University Press, 2016)

Marek Cejka, 'Divide et Impera?: Western engagement in the Middle East' , *Central European journal of International and Security Studies*, Vol 6, Issue 4(2012) pp.200-219

Jonathan N.C. Hill, 'Identity and instability in postcolonial Algeria', *the Journal of North African Studies* 11:1, (2006), 1-16

Sara Pursley, *'Lines Drawn in the Sand'*, Jadaliyya
https://www.jadaliyya.com/Details/32140 [Part 1] jun 2, 2015
https://www.jadaliyya.com/Details/32153 [Part 2] accessed 6 march 2020

Daniel Neep, *'FOCUS: THE MIDDLE EAST, HALLUCINATION, AND THE CARTOGRAPHIC IMAGINATION'*
https://discoversociety.org/2015/01/03/focus-the-middle-east-hallucination-and-the-cartographic-imagination/ accessed 6 march 2020

THE EUROPEAN INTEGRATION PROCESS FOR MAINTAINING PEACE: SUCCESSES AND FAILURES

The geographical area of Europe has seen many wars in history. Since the creation of the firsts societies, men found thousands of reasons for starting wars against each other.

However, there is an equal number of reasons for not starting those conflicts that put people at risk but the condition to make this possible are challenging to achieve and to maintain. Europe might be the most exciting case in the world. It could be argued that the countries that made up Europe are some of the most belligerent in the world. A country such as the United Kingdom, France and Germany have been strongly expansionist in their foreign policy.

The significant majority of European country has sought, in a certain moment in history, not only to subjugate other parts of the world through colonisation, but to subdue their neighbours using coercive methods. It is no coincidence if two of the worst conflicts that history has ever seen took place because of European countries and then expanding all over the world involving other territories. After the second world war, all this gradually changed.

Except for some war occurred during the decolonisation process, the European countries, the most aggressive nations in the world, lost a significant part of their violent attitude in foreign policy, especially in the relationship among other European countries. The purpose of this essay is to explain what happened during the process of the strengthen interdependence that made Europe more resilient. Furthermore, it will be argued that while the process of European integration guaranteed a peaceful framework among its members, it also struggled in preventing war outside its boundaries but still on its continent. There will be a short analysis about the wars in the Balkans during the 90s in which the European Union did little to improve the situation because it lacked of the right instruments to act coherently.

The final argument of the essay will be that economic integration, can lead to peace but also to some dangerous drawbacks, especially if the countries economically integrated did not develop a coherent supernational approach to some issues such as foreign policy and defense, showing that in some cases, economic integration is not enough to avoid conflicts.

In order to properly understand all these issues, it is necessary to start explaining what prompted the European integration project by giving a general historical context.

After the second world war, Europe was at the edge of the abyss. The European merchant fleets were destroyed, there were no more foreign investments, and the foreign markets where the Europeans used to sell their products were infiltrated by goods produced in the United States, Canada and underdeveloped countries that took advantage of the war to replace European industries. Millions of people were threatened by hunger and poverty in both winning and losing countries.[92]

The new international situation and the defeat of the nazi-fascism contributed to make the political elite aware of the necessity to build an international framework of cooperation. Maintaining peace was a primary objective at that time. The attempt to avoid war during the second half of the 30s failed miserably and, as a consequence, the way in which this peace was preserved had to change drastically using new institutions, a strengthen intergovernmental system and a different European order.

A first proposal to found a "United Europe" can be identified in the *Ventotene Manifesto* published in 1941 by Altiero Spinelli, a blueprint for a federation of European states. According to Spinelli, the reason why Europe struggled in finding a peaceful framework in which operate was that the international system was anarchic.

---

[92] R. Cameron, L. Neal, *Storia economica del mondo, vol. II Dal XVIII secolo ai nostri giorni*, (Bologna 2005) pp. 581-582

This means that every country pursues its individual interests rather than a communal one. The only way to change this system is the formation of new institutions capable of imposing rules over a group of nations. However, this was a theory already formulated by multiple philosopher and politicians. When Spinelli was writing his Manifesto, Europe was in the middle of a tragic war and Nazi Germany was trying to establish its own international order based on violence and coercion like many in Europe already tried in the past. Every similar attempt failed in history.

The creation of a single European State, precisely in the Empire conformation, was something impossible and too violent. Spinelli argues that a more suitable option would have been a federation. Basically, Spinelli advocated for the United States of Europe with a common army, the capability of choosing its foreign policy without States' intromission, the abolition of trade barriers and the currency creation. Spinelli was visionary. Unfortunately, his theory was impossible to achieve in the short term.[93]

Nevertheless, something started to move in the direction of more cooperation between States. Belgium, Luxemburg and the Netherlands established the *Benelux*, a customs union among their countries in order to foster their economies in the reconstruction process during the post-war period.

---

[93] Altiero Spinelli, *Il Manifesto di Ventotene*, 1941

At the same time, the United Kingdom and France signed the Treaty of Dunkirk establishing a defensive alliance and a mutual assistance pact in order to protect themselves from a German military revival. With the Treaty of Brussels also the Benelux joined. Luckily, this initiative will be proved as useless because the United States soon realised that the new world order could not be established without their participation. As a result, they successfully withdraw from their isolationist foreign policy moving into full participation in defence of Europe signing and promoting the North Atlantic Treaty Organization in which every country would have contributed to a military and even political cooperation.

However, the European project of integration was not shared by every country. There were different ideas on what kind of Europe it was necessary to build.

On the one hand, there was a more supernational project supported by France, Benelux and Italy. On the other hand, the United Kingdom, Switzerland and the Scandinavian countries preferred a more intergovernmental cooperation.[94]

It could be argued that these two different models still divide Europeans today, and the contemporary European Union is a mix of these two approaches.

---

[94] Cini, M. and Borragàn, *European Union Politics*, (Oxford, 2019) p. 14

The supernational approach to develop a peaceful environment for Europe is something that traces its roots to the eighteenth century, during the Enlightenment, when Kant theorised some preconditions for guaranteeing perpetual peace. According to the German philosopher, one of these preconditions is to a have a federation of republics. In other words, a supernational entity.[95]

This issue will be mentioned again in the following sections of the research paper, when it will be considered the way in which the European Union responded to some challenges.

Going back to the foundation of Europe, in 1948, some West European countries founded the Organisation for European Economic Cooperation. Its purpose was to coordinate U.S. financial assistance from the Marshall Plan and to encourage trade liberalisation.

In 1949, Belgium, Denmark, France, Ireland, Italy, Luxemburg, the Netherlands, Norway, Sweden and the United Kingdom established the Council of Europe. This was a significant first step toward integration, even if the Council was just an intergovernmental institution.

The revolutionary breakthrough was taken the following year. In 1950, the French Foreign Minister, Robert Schuman, presented a declaration which would have gone down in history.

---

[95] Immanuel Kant, *per la pace perpetua*, (Milano, 2004) pp. 61-66

Basically, Schuman stated that his plan was to proceed to a sectoral integration of coal and steel resources between France and Germany, also inviting other European countries to participate. This initiative would have made <<any war between France and Germany [...] not merely unthinkable but materially impossible.>>[96] This allowed the Six founding members, France, Germany, Belgium, Netherlands, Luxembourg and Italy to create the first supernational institutions: the European Coal and Steel Community.

At this point, it is evident that the economic integration and the promotion of interdependence among countries was considered the precondition of peace by the founding fathers of Europe.

Basically, this was the beginning of a long process with many steps and evolutions. It would be pedantic summarise every event, treaty or conference that took place in history. As a consequence, the focus will be now put on some thesis of the European integration.

According to Alan Milward and to Andrew Moravcsick, the European integration depended significantly to the willingness of the Six founders to reach prosperity, internal security regarding Germany and external security regarding the communist threat posed by the USSR.

---

[96] Robert Schuman, *Declaration,* 9 May 1950

Moravcsick emphasises, even more, the economic reasons that prompted the integration by focusing his attention to the trade issues.[97]

On the other hand, Mark Gilbert argues that the economy was without any doubt relevant, but it is crucial also to underline political factors. In the aftermath of the second world war, the political elites in Europe embraced a new political idealism in which nationalist interests were put aside in favour of community initiatives. When in 1950 Germany was about to surpass France industrial production, the French foreign minister could have reacted by diplomatically isolating Germany or by adopting protectionist policies. On the contrary, five years later the end of the most violent war in history, political elites in both countries chose to cooperate. Adenauer and Schuman thought this was the only solution for European safety. According to Mark Gilbert this can be only a demonstration of idealism.[98]

Despite that, idealism was not enough for proceeding without obstacles. The political elites in Europe drastically changed their way of thinking after the second world war, but they had some boundaries that could not be crossed easily. The European Coal and Steel Community was one of these boundaries.

---

[97] Mark Gilbert, *Storia politica dell'integrazione europea*, (Roma-Bari 2017) p. X
[98] Mark Gilbert, *Storia politica dell'integrazione europea*, (Roma-Bari 2017) p. XI

Beyond the ECSC there would have been further integration toward a more supernational organisation by building the European Defense Community, a common European army that would have costed too much sovereignty to the European countries. As a result, the project was not ratified by the French parliament, and the integration process had its first rollback. The truth is that the EDC was the European response to the United States' request to re-arm Germany. In 1950, Washington considered the rearmament a priority because the Korean war looked like a scenario that could have also verified in Europe. There was a scary parallelism between Korea and Germany. The USA wanted to make sure to have an efficient structure of defence in case of war in Europe. Even if the EDC would have been a European military organisation, article 18 stated that the supreme commander would have been responsible to the NATO.[99]

Nevertheless, the project failed to be applied, and a common European foreign policy was not organised at that point in history. NATO was the only way for Europeans to defend themselves against external threats and the only military framework in which cooperate.

This would have had consequences during the 90s.

---

[99] European Defense Community Treaty, 27 May 1952 https://aei.pitt.edu/5201/1/5201.pdf [Accessed 11/05/2020]

As a result of the failure in integrating European countries through a truly supernational organisation that would have included a common army and a political body able to make decisions about foreign policy, the newborn European organisation, the European Economic Community, established its legitimacy focusing on the economic field rather than the political one. As a demonstration, in 1979 was established a European Monetary System which purpose was to increase the cooperation between the Central Banks and to stabilise the exchange rate. This would have contributed to make trade easier among European countries.[100]

To sum up the integration process during the years, the 50s proved to be the years in which the idea of Europe started to build a common project on the ashes of the second world war. During the 60s, the French president De Gaulle tried to establish his own view of Europe based on intergovernmental approach,[101] while the 70s were the decade in which the Community enlarged itself and increased the level of economic cooperation with some lukewarm attempt to deepen the political cooperation on foreign policy with the Werner Report on Economic and Monetary Union.[102]

---

[100] Cini, M. and Borragàn, *European Union Politics*, (Oxford, 2019) p. 18
[101] Cini, M. and Borragàn, *European Union Politics*, (Oxford, 2019) p. 16
[102] Cini, M. and Borragàn, *European Union Politics*, (Oxford, 2019) pp.16-18

The 80s were the decade in which a new wave of support for European integration hit political elites. Both the French president Mitterand and the German Chancellor Kohl demonstrated their commitment to make some further steps toward a unified Europe.[103] The turning point of this renewed spirit was the signing of the Single European Act in 1993. Its purpose was to form the European Single Market, an area in which Europeans could move goods, capitals, services and labour freely. Additionally, the SEA would have fostered the transformation of the Community into a Union introducing new competencies such as the environment, research and development, economic and social cohesion and so on. Moreover, the reform significantly empowered the European Parliament through the implementation of the ordinary legislative procedure, the way in which the parliament can affect the legislative process. Last but not least, the qualified majority vote in the Council expended covering many issues. [104]

Most importantly, the treaty on European Union signed in Maastricht introduced also a common foreign policy and security matters and justice and home affairs.

---

[103] Helmut Kohl and François *Mitterand: leaders in Reconciliation, European Union official website,*https://europa.eu/europeanunion/sites/europaeu/files/eu_pioneers_kohl_mitterrand_en.pdf[Accessed 11/05/2020]

[13] Cini, M. and Borragàn, *European Union Politics,* (Oxford 2019), p. 20

Those implementations were the consequence of some international changes in the European framework that will be analysed in the next section of the research paper. However, these two issues were based on intergovernmental cooperation. Deirdre Curtin defined this new European structure as a <<Europe of bits and pieces>>[105] because in his opinion, it was an incoherent mix of supernational and intergovernmental characteristics.

After having described as shortly as possible, the *genesis* of the European Union and its main problems, it will be necessary now, turn the attention to the first truly violent crisis that Europe faced after the second world war. During the second half of the 20th century, there have been at least two violent actions in Eastern Europe that threatened the security of the country. The Soviet invasion of Budapest and the Warsaw Pact invasion of Prague. Both succeeded in keeping the *status quo* without leading to a continental war. In addition, there have been some cases of high tension between the west and east around the German issue during the cold war, but none of them led to a war. The first official large conflict in Europe after the second world war took place in the Balkans; a region called the *powder keg of Europe* since the dissolution of the ottoman empire because of the number of conflicts that occurred in that region.

---

[105] Deirdre Curtin, "The Constitutional Structure of the Union: A Europe of Bits and Pieces." *Common Market Law Review 30.1* (1993), pp. 17-69

The second world war was not only the crucial moment that triggered the process of European integration. In the Balkans, out of this war, was founded a Yugoslavian federation for the second time, made up of six republics: Croatia, Montenegro, Serbia, Slovenia, Macedonia, Bosnia and Herzegovina. Josip Broz Tito, led the biggest communist-partisan movement during the second world war. They independently fought the Nazis, and they freed their country from the invaders and as a consequence, the country resulted strictly controlled by the communists and his leader, Tito.[106] It ought to be also said that each republic had an ethnically heterogeneous population within its border. This would have led to severe consequences during the dissolution of the Yugoslavian federation because the coexistence among different ethnicities hid some terrible hostilities held off only with the effort of Tito's communist leadership and Serbian hegemony.

This fragile balance based on communist control and Serbian predominance was lost after Tito's death in 1880. His charisma contributed to keep the unity of the federation. However, when he died, his follower Milosevic, tried to consolidate his power by strengthening Serbian control over Kosovo, a small region of Serbia that aimed to achieve more autonomy.

---

[106] John Alcock, John Lampe, Yugoslavia, Former federated nation 1929-2003, *Encyclopedia Britannica*, 2019

Simultaneously also in the other republics grew some dissatisfaction with the constitutional setting. Milosevic decided to act aggressively to the raising opposition in 1989 by revoking the autonomy of Kosovo and Vojvodina and imposing martial law. This was the drop that broke the camel's back. The following year, Slovenian delegates to the Extraordinary Congress of the League of Communists asked for a multi-party system. In Serbian communist's vision, this was impossible to concede, while the republics of Croatia and Bosnia and Herzegovina quickly decided to stand for more autonomy by supporting Slovenian requests. [107]

Paradoxically, in the years in which European integration was consolidating the European Union, even if with some serious problems, the Yugoslavian integration project was drastically falling apart.

Having introduced both the European integration process and the beginning of the dissolution of Yugoslavia, let us turn the attention on what kind of choices the European Community made in order to deal with the situation.

---

[107] Annex IV: The policy of ethnic cleansing, *Final report of the United Nations Commission of Experts established pursuant to security council resolution 780* (1992), https://web.archive.org/web/20120504142243/http://www.ess.uwe.ac.uk/comexpert/ANX/IV.htm#0-IV , [Accessed 27/05/2020]

Firstly, the break-up of the federation was not impossible to foresee. George Kennan, stated in 1989: <<events in Yugoslavia are going to turn violent and confronting Western countries, especially the United States, with one of their biggest foreign policy problems in the next few years.>>[108]

However, at that time, Washington was focused on the dissolution of USSR and Yugoslavia was a minor problem to deal with.

Secondly, in 1989, the E.C. was convinced that by financially supporting the Yugoslavian reforms, the federation could become something similar to the E.C., and this would have avoided a violent conflict.[109]

With the United States unwilling to act in that region and the United Nations' Secretary-General arguing that he could have any role in the matter because <<Slovenia [was] not an independent U.N. member>>[110], the E.C. was supposed to take a leading position.

---

[108] Louise Branson, Dusko Doder, *Milosevic: portrait of a tyrant*, (New York, 1999), chapter 5

[109] Sonia Lucarelli, *Western Europe and the Breakup of Yugoslavia A political failure in search of a scholarly explanation*, (Florence, 1998) p.13

[110] Sonia Lucarelli, *Western Europe and the Breakup of Yugoslavia*, (Florence, 1998) p. 15

On 25 June 1991, Slovenia and Croatia declared their independence. When the war started between Slovenia and Serbia, the E.C. stopped its financial support to Yugoslavia while also activating the Organization for security and Cooperation in Europe. The plan was to act as a mediator and to resolve the issue as soon as possible. Fortunately, the Troika diplomatic mission put in place by the E.C. succeeded in negotiating the Brioni Agreement which would have suspended the declaration of independence and stopped Serbian invasion. While in Slovenia the war lasted only ten days thanks to the E.C. intervention, in Croatia the situation proved to be more difficult because the E.C.'s diplomatic efforts focused mainly on Slovenia rather than Croatia[111].

The Serbian minority living in Croatia but supported by Serbia, soon declared independence for their region in Krajina. When the conflict expanded there, the French foreign minister proposed to send a lightly-armed WEU peace-keeping force in Yugoslavia but some countries as Britain, Spain, Greece and Germany were unwilling to intervene militarily.[112]

There are two considerations worthy of being made.

---

[111] Sonia Lucarelli, *Western Europe and the Breakup of Yugoslavia*, (Florence, 1998) p. 16

[112] Sonia Lucarelli, *Western Europe and the Breakup of Yugoslavia*, (Florence, 1998), p.17

Firstly, it has been already explained the way in which the project of a European army failed in the past. That integration process was stopped in the 50s and, as a result, when it came to be a necessity, the European Community did not have the right instrument to act. In this case, European integration was not enough to prevent war.

Secondly, the economic and political integration of the republics that formed Yugoslavia in a federation, similar to what the wiser supporters of federalism in Europe dreamed, was not enough to stop the outbreak of a war.

The solution could only be diplomatic for the E.C. Therefore, the E.C. sponsored the London Conference to mediate among the six republics of Yugoslavia. However, the attempt failed due to many contrasts among the European countries that were not able to impose their condition to Serbia that was becoming increasingly hostile toward the E.C. mediation effort. The difficulties in achieving some remarkable result, persuaded France in bringing the U.N. in. Even with the participation of the U.N., the main international actor still was the EC but some countries, like France, hoped for a U.N. peace-keeping mission as the only way to guarantee a cease-fire agreement. In the meantime, Germany, Italy and Denmark supported the recognition of Slovenia and Croatia as new states. On the contrary, France and other E.C. members opposed to this recognition.

This only proved the weakness of the E.C. in achieving the objective of peace guarantor because of its difficulties in taking a firm position on its foreign policy. Eventually, Germany decided to stand for the recognition of Slovenia and Croatia. Slowly all the E.C. members did the same while the tension in Bosnia and Herzegovina drastically increased.[113]

In April 1992, also Bosnia and Herzegovina followed the same path declaring independence and starting a war against Serbia and the Serbian minority in the republic. At this point, the international community and the European Community recognised the new republics and asked for Serbian withdrawn. However, in Bosnia and Herzegovina the JNA, the Serbian army and the ethnic minority, decided to start a genocide and an ethnic cleansing against the Bosnian population. After almost 50 years of peace in Europe, a terrible disaster was happening again. Mass killing, concentration camps, confiscations, looting and rapes that the European integration project seemed to have eradicated from the continent, was soon taking back control of the foreign policies of one state.[114]

---

[113] Sonia Lucarelli, *Western Europe and the Breakup of Yugoslavia*, (Florence, 1998), pp. 18-22

[114] Annex IV: The policy of ethnic cleansing, Final report of the United Nations Commission of Experts established pursuant to security council resolution 780 (1992), https://web.archive.org/web/20120504142243/http://www.ess.uwe.ac.uk/comexpert/ANX/IV.htm#0-IV , [Accessed 27/05/2020]

In order to have an idea of the size of the conflict, it could be useful to point out some figures. In Bosnia and Herzegovina, the number of casualties has been esteemed between 89.000 and 100.000. Approximately the 65% of them were Bosniak Muslims while the 20% were Serbs and of the remaining there were Croats and other ethnicities. [115]

It was logical to think that the dark memories of the second world war and of the Nazi's concentration camps should have prevented this carnage, but this was not the case. Moreover, the conflict intensified when Croatians living in Bosnia and Herzegovina also decided to fight for annexing their territories to the bordering Croatia.

The E.C. put to Serbia a deadline for the withdrawn of its forces in Bosnia and Herzegovina, but this did not have any effect. The only action that the E.C. could take was to deploy humanitarian aids while its member states still hoped for the U.N. intervention. On their part, the U.N. and USA advocated for a European management of the conflict.

Meanwhile, another issue emerged. The war caused the exodus of 1.4 million of refugees.

---

[115] Jan Zwierzchowski, Ewa Tabeau, *The 1992-95 war in Bosnia and Herzegovina: census-based multiple system estimation of casualties undercount*, in Conference Paper for the International Research Workshopon 'The Global Costs of Conflict' The Households in Conflict Network (HiCN) and The German Institute for Economic Research (DIW Berlin), (Berlin, 2010)

Germany hosted the majority of them, and the E.C. members were unwilling to emulate the example.[116] The war refugees will be a constant reason of friction among Europeans in many more occasions.

In the following months, the international press started to publish the atrocities of the ethnical cleansing at the expenses of the Bosnian and of the Muslims living those regions. Despite the catastrophic situation, the E.C. and the international community did no more than sending humanitarian aid, establishing "no-fly zones" and imposing sanctions and restrictions to the newly created Federal Republic of Yugoslavia, formed with the remaining two republics of Serbia and Montenegro. [117]

It was precisely the "no-fly zone" that prompted NATO intervention in 1993. The FRY ignored the imposition, and as a consequence, NATO intervened with the purpose of engaging any hostile Serbian aircraft. It has been already explained that the lack of a European army could cause some problems. In fact, the only way to cooperate in the military field for Europeans was to rely on NATO and the U.S. involvement.

---

[116] Joelle Hageboutros, "The Bosnian Refugee Crisis: A Comparative Study of German and Austrian Reactions and Responses." *Swarthmore International Relations Journal* Iss. 1 (2016): 50-60

[117] Sonia Lucarelli, *Western Europe and the Breakup of Yugoslavia*, (Florence, 1998), p.33

However, in this case, the U.S. administration was initially unwilling to take a significant role in the conflict. [118]

A further weak diplomatic attempt was made with the Vance-Owen Plan, a joint U.S. and E.C. effort to peacefully resolve the situation by dividing the country into autonomous provinces in order to separate the different populations ethnically. Nevertheless, the Serbs did not like the plan because it would have reduced the territory under their control, and as a result, by the end of 1993 the plan was already wrecked.[119] The clear rejection caused the continuation of the conflict with the death of civilians and innocent people, especially in Sarajevo, where in 1994 a marketplace was bombed with 68 casualties.[120]

The war reached a point of no return, and NATO was obliged to make an ultimatum: the Serbian forces should have withdrawn in ten days, but they refused and continued to bomb U.N.'s safe zones and sensitive areas such as hospitals. This time NATO responded with an airstrike of Bosnian Serb command post.

---

[118] Sonia Lucarelli, *Western Europe and the Breakup of Yugoslavia*, (Florence, 1998), p.40
[119] The Vance-Owen Plan, https://www.peaceagreements.org/wview/606/The%20Vance-Owen%20Plan [Accessed 25/05/2020]
[120] Jim Fish, 'Sarajevo massacre remembered' , *BBC*, February 2004

The violent escalation was unstoppable, and the Serbs increased their atrocities while NATO boosted the bombing campaign.[121]

Meanwhile, in May and August 1995, Croatia prepared its offensive against its original territories now controlled by the Serbs due to the sizeable Serb population living there. At this time, Serbs were obliged to fight on multiple front and enemies. This caused many difficulties and put them in the condition to accept a peace treaty.

Eventually, the USA sponsored some talks in Dayton in which a new and definitive agreement was reached between the parties. The Dayton Agreement consisted in the separation of Bosnia and Herzegovina in two entities: one is the Federation of Bosnia, and Herzegovina with a Bosniaks and Croats majority and the other one is the Republic of Srpska with a majority of Serbs. [122]

The war was not over. In 1999, Kosovo started its own fight against Serbia in order to achieve what the other republics already won. However, the focus of this research paper is on the first and main part of the Yugoslavian conflict that has been explained above.

---

[121]Ryan C. Hendrickson, *Crossing the Rubicon*, 2005
https://www.nato.int/docu/review/2005/issue3/english/history_pr.html [Accessed 25/05/2020]

[122] Filippo Andreatta, 'The Bosnian War and the New World Order: Failure and Success of International Intervention'

Institute for Security Studies, Western European Union, 1997

During this phase of the conflict, between 1990 and 1995, it has been undoubtedly proved that the E.C. was not able to maintain peace in Europe outside of its border.

The only way to achieve such a result was involving the U.N. and NATO; therefore, Washington. The American leadership proved to be essential for the European stability since the second world war. Even if the European integration process made great strides, still common decision was difficult to achieve, proving that economic integration at that point was not sufficient to guarantee peace. These events contributed to make the E.C. member states aware of the limit of the community, also fostering, for some extent, a process of significant integration in a more unionist project during the 90s.

Additionally, the evidence presented, demonstrates that economic integration does not always prevent war.

Firstly, in the case of Yugoslavia, the federation was torn apart by conflicts even if it was economically well integrated. It cannot be ignored that there were many reasons why the war broke out, but still the economic ties among the Yugoslavian republics did little to prevent the conflict.

Secondly, in the case of the E.C., even if the community built a peaceful framework for western European countries for years, economic integration was not enough when it came to act coherently and unitedly.

Political situation in 2010

OOO - Bosnia and Herzegovina   O - Serbia

A possible explanation and interpretation is that in some cases, economic interdependence is not enough to prevent war. A mix of supernational and intergovernmental structures could prove useless in some contexts, especially if the E.C. need the intervention of NATO when it could deal with the situation by itself. Many observers claim that it would be necessary for the E.U. to achieve more political cooperation.

In 1990 John Mearsheimer tried to predict the future. He wrote an article in which, using a neorealist theory to understand the post-Cold War European situation, he argued that Europe would have returned to a dangerous situation of anarchy. In his article "Back to the Future", the changed international political scene would have caused the failure of international institutions to maintain peace. According to Mearsheimer, Europe was predestinated to a future of new conflicts.[123] Unsurprisingly Robert Keohane, Joseph Nye and Stanley Hoffmann opposed to such a theory arguing that it was the highly institutionalization and cooperation in Western Europe that would have proved to be essential in the process of stabilization of the continent. [124]

---

[123] John J Mearsheimer., Back to the Future: Instability in Europe after the Cold War, *International Security*, Vol. 15, No. 1. (Summer, 1990), pp. 5-56

[124] Sonia Lucarelli, *Western Europe and the Breakup of Yugoslavia*, (Florence, 1998) p.3

Keohane and Nye were among the first political scientists that noted the change in the global framework where the paradigm of state centrism lost predominance for understanding transnational relations.[125]

In conclusion, both the theories were partially correct. Anarchy did not prevail because EU managed to survive the cold war and it even enlarged itself during the 90s and early 2000s. Nevertheless, it cannot be argued that the EU completely succeed in keeping peace in Europe because its institutions, even if integrated, allowed a terrible war to happen in the continent, showing all its limits face to the challenge of instability.

---

[125] Keohane, Robert, Nye, Joseph, 'Power and Interdependence Revisited', International Organization, 41, 4 (1987), 725-53

Bibliography & sitography

Alcock John, Lampe John, Yugoslavia, Former federated nation 1929-2003, Encyclopedia Britannica, 2019 https://www.britannica.com/place/Yugoslavia-former-federated-nation-1929-2003, [Accessed 27/05/2020]

Andreatta Filippo, 'The Bosnian War and the New World Order: Failure and Success of International Intervention' *Institute for Security Studies, Western European Union*, 1997

Annex IV: The policy of ethnic cleansing, Final report of the United Nations Commission of Experts established pursuant to security council resolution 780 (1992), https://web.archive.org/web/20120504142243/http://www.ess.uwe.ac.uk/comexpert/ANX/IV.htm#0-IV , [Accessed 27/05/2020]

Branson Louise, Doder Dusko, *Milosevic: portrait of a tyrant*, New York, 1999

Cameron, Rondo, Neal, Larry, *Storia economica del mondo, vol. II, Dal XVIII secolo ai nostri giorni*, Bologna, 2005

Cini, M. and Borragàn, *European Union Politics*, Oxford University Press, N.P.S. 2019

Curtin, Deirdre, "The Constitutional Structure of the Union: A Europe of Bits and Pieces." *Common Market Law Review* 30.1 (1993): pp. 17-69

European Defense Community Treaty, 27 May 1952 https://aei.pitt.edu/5201/1/5201.pdf [Accessed 11/05/2020]

Fish Jim, 'Sarajevo massacre remembered', *BBC*, Februry 2004

Gilbert Mark, *Storia politica dell'integrazione europea*, Roma-Bari, 2017

Hageboutros, Joelle. "The Bosnian Refugee Crisis: A Comparative Study of German and Austrian Reactions and Responses." *Swarthmore International Relations Journal* Iss. 1 (2016): 50-60

Helmut Kohl and François *Mitterand: leaders in Reconciliation, European Union official website* https://europa.eu/european-union/sites/europaeu/files/eu_pioneers_kohl_mitterrand_en.pdf [Accessed 11/05/2020]

Hendrickson Ryan C., *Crossing the Rubicon*, 2005

https://www.nato.int/docu/review/2005/issue3/english/history_pr.html [Accessed 25/05/2020]

John J Mearsheimer, Back to the Future: Instability in Europe after the Cold War, *International Security*, Vol. 15, No. 1. (Summer, 1990), pp. 5-56

Keohane, Robert, Nye, Joseph, 'Power and Interdependence Revisited', *International Organization*, 41, 4 , 1987, 725-53
Lucarelli Sonia, *Western Europe and the Breakup of Yugoslavia A political failure in search of a scholarly explanation*, Florence, 1998

Schuman Robert, Declaration, 9 1950, https://europa.eu/european-union/about-eu/symbols/europe-day/schuman-declaration_en [accessed 11/05/2020]Spinelli Altiero, Il Manifesto di Ventotene, 1941

http://www.altierospinelli.org/manifesto/it/statiunitiit_en.html [Accessed 27/05/2020]

The Vance-Owen Plan, https://www.peaceagreements.org/wview/606/The%20Vance-Owen%20Plan [Accessed 25/05/2020]

Zwierzchowski Jan, Tabeau Ewa, *The 1992-95 war in Bosnia and Herzegovina: census-based multiple system estimation of casualties undercount*, in Conference Paper for the International Research Workshopon 'The Global Costs of Conflict' The Households in Conflict Network (HiCN) and The German Institute for Economic Research (DIW Berlin), Berlin,

2010 https://www.icty.org/x/file/About/OTP/War_Demographics/en/bih_casualty_undercount_conf_paper_100201.pdf [Accessed 27/05/2020]

WHAT WENT WRONG WITH TURKEY INTEGRATION IN THE EUROPEAN UNION

During European Union's history the process of enlargement has seen different stages in which new candidates progressively joined the union. Most of them just spent few years between their candidature and their approval while Turkey is the longest-standing EU candidate so far.[126] My argument is that one of the main reasons that prevented Turkey from joining EU is that other member states claim a lack of "Europeanness" in Turkey's culture.

But is that really a good argument for excluding this country?

European Union and Turkey relationship dates back to 1964 when the European Economic Community and Turkey signed an Association agreement. In 1987 the Turkish government applied for becoming a member state while some years later, in 1996, the two established successfully a Customs Union. But it was only in 2005 that the European Union opened the negotiation for a further integration. Twelve years later, in July 2017, the European Parliament decided to suspend the negotiation. [127]

---

[126] Cini, M. and Borragàn, *European Union Politics*, (Oxford University Press), N.P.S. 2019, 270.
[127] Cini, M. and Borragàn, *European Union Politics*, 270.

On one hand, some people argue that Turkey is culturally incompatible with European Union because of its past history and Islamic values as the perfect depiction of otherness regarding European culture.[128] On the other hand, some argues that this is not necessarily true because Turkish culture is not completely irreconcilable with European culture.

After 9/11, Huntington's *Clash of Civilization* came up as the theory whose purpose is to interpret the new international world order after the Cold War. The hostility between Christianity and Islam emerged again while European Union was beginning its negotiation with Turkey. As a consequence, the Europeanness of Turkey has been questioned opening the debate about the role of Christianity in the European identity.

While some argue that Christian heritage is a fundamental characteristic of all European countries, other claim that European Union is a secular institution in which religion should not matter anymore.[129]

---

[128] Kristy Hughes, Turkey and the European Union: just another enlargement? Exploring the implications of Turkish accession, pp. 1-2

[129] Dietrich Jung, Catherina Rauvere, *Religions, politics and Turkey's accession*, (Palgrave Macmillan 2008), 7.

However, Nicholas Sarkozy during a speech as a candidate for the presidency stated that Turkey "does not have its place" in EU and that the French are the "heirs of 2000 years of Christianity".[130]

In addition, both French and Austrian government announced that they would call a public referendum for Turkey's membership. This remarked their firm opposition to Turkey integration because of its lack of Christianity.[131]

During an interview with *Le Monde*, Valéry Giscard d'Estaing argued that geographically speaking Turkey is not even in Europe as its capital and the 95% of its population is in Asia Minor. Despite d'Estaign' s statement, European boundaries have never been unanimously identified.[132]

Furthermore, Turkish territory and its population are a consequence of specific historic developments occurred after the first world war which shrank Turkish lands pulling out the former Ottoman Empire from what is now considered Europe. In fact, for centuries the Ottoman Empire has been intertwined with other European kingdoms during both war and peaceful times.

---

[130] Dietrich Jung, Catherina Rauvere, *Religions*, 29.
[131] H. Tarik Oğuzlu Turkey and the European Union: Europeanization Without Membership, *Turkish Studies*, 13:2, (2012), 234
[132] Dietrich Jung, Catherina Rauvere, *Religions* 8.

For instance, the French kingdom and the Ottoman empire signed an alliance in 1535 which lasted until the nineteenth century and subsequently Great Britain supported the Ottomans in order to contain Russian expansionism.

In addition, Ottomans have started in the nineteenth century a process of reformation, known as Tanzimat, whose purpose was to modernize the country and the outcome was similar to the process already started by some Europeans country.

When the empire collapsed, the new Turkish Republic pursued its policies of modernization sharing a similar path with Europe. Moreover, during the twentieth century Turkey cooperated with Europe in multiple ways, from the economic to the military point of view and eventually joining NATO.

Despite these elements, according to a 2005 Eurostat survey, less than 30% of the public opinion is positive toward Turkey integration.[133] Furthermore, increasing support for right-wing populist parties in the last parliamentary elections is fostering hostility against migrants and Muslims in European countries.[134]

---

[133] Dietrich Jung, Catherina Rauvere, *Religions*, 28.
[134] Cini, M. and Borragàn, *European Union Politics*, 270

As demonstration of Turkey's closeness to Europe, some political parties like the CHP, the NAP and especially the AKP have shown for some extent and with different degrees, willingness to join Europe because in their opinion this would led to further democratization, westernization and Europeanization.[135]

Politically speaking, Turkey has been through a process of reformation not comparable with the one in which eastern European countries have gone through but still this process has involved Turkey's political life in making it more democratic. Despite Turkey's democratization is not yet complete and it is facing some serious problems, the attempt to transform and secularize the country is the demonstration of what kind of efforts Turkey has been facing in order to be recognized as part of Europe.[136] If Europe could overcome the challenge of integrating what has been considered as "the permanent other" not by assimilation but through a really multiculturalist approach, this would result in a great achievement which could turn up useful for integrating Muslim minority living in the European member states as well.[137]

---

[135] H. Tarik Oğuzlu Turkey and the European Union: Europeanization Without Membership, *Turkish Studies*, 13:2, (2012), 233

[136] Kristy Hughes, Turkey and the European Union: just another enlargement? Exploring the implications of Turkish accession, p. 3

[137] Hasan Kosebalaban, The Permanent "Other?" Turkey and the question of European Identity, *Mediterranean Quarterly* 18, no. 4, (2007): 87-111

The point is that culture does not seem as a sufficient reason for excluding Turkey from the European Union. There are undoubtedly many things that went wrong during Turkish integration and I suggest that one of these reasons is the hostility of some European actors in admitting that Turkey could merge with European culture.

Perhaps, rather than focusing on the cultural issue, it would be more profitable pay attention whether Turkey will become a fully democratic country ready to join the union or not.

Bibliography

Cini, M. and Borragàn, *European Union Politics*, Oxford University Press, N.P.S. 2019

Dietrich Jung, Catherina Rauvere, *Religions, politics and Turkey's accession,* Palgrave Macmillan 2008

Hasan Kosebalaban, The Permanent "Other?" Turkey and the question of European Identity, *Mediterranean Quarterly* 18, no. 4, (2007): pp. 87-111 https://www.muse.jhu.edu/article/224690 [Accessed 26 February 2020]

H. Tarik Oğuzlu Turkey and the European Union: Europeanization Without Membership, *Turkish Studies*, 13:2, (2012) pp. 229-243 https://doi.org/10.1080/14683849.2012.685256

Kristy Hughes, Turkey and the European Union: just another enlargement? Exploring the implications of Turkish accession, EuroActive.com, June 2004, https://www.euractiv.com/section/enlargement/opinion/turkey-and-the-european-union-just-another-enlargement/ [Accessed 28 February 2020]

CONCLUSIONS

This book is an attempt to reconstruct a small part of European history.

 The incipit is placed during the beginning of the age of empires, when explorations and discoveries put the Europeans in front of alien cultures. Surely, this was not the first time in which different cultures came in contact with something different from them. However, it was the first time in history where a so dramatic encounter took place.

In this collection of academic papers, there are not singular histories of small events, on the contrary, this is the collection of singular moment in histories that fill perfectly the bigger picture of a great European adventure in history.

During this spectacular voyage Europe discovered the world and then, with its own instruments and tools learned how to conquer new territories and people beyond its own imagination. In the process of discovering the *other*, Europe also depicted itself by acquiring a self-portrayed identity. Soon, this identity turned out to be despotic, violent, predatory and aggressive.

 Science, explorations, economics and every European feature was implemented in the astonishing effort of subjugate those alterities that Europe discovered outside its borders.

The consequences of this events were disastrous. Many countries were left to deal with Europeans actions. The same old continent risked to implode under a destructive war.

However, every cloud has a silver lining and from the ashes of such destruction, Europe managed to build peace.

Yet, the challenge has not been overcome completely. Recent events as the conflict in the Balkans during the 90s and the attempt of Turkey integration and suddenly downgrading in a less democratic state, posed new problems to Europe, demonstrating that what has been constructed during the years after the Second World War is not an easy task to maintain.

If the European Union wants to survive and evolve in something more stable for its internal and external peace, the path is easy to be distinguished but difficult to follow.

Democratic integration is the only sustainable solution.

Lightning Source UK Ltd.
Milton Keynes UK
UKHW022110110621
385375UK00002B/238